Sensors in Bioprocess Control

Bioprocess Technology

Series Editor

W. Courtney McGregor

Xoma Corporation
Berkeley, California

Sensors in Bioprocess Control

edited by

John V. Twork
Tennessee Eastman Company
Kingsport, Tennessee

Alexander M. Yacynych
Rutgers, The State University of New Jersey
New Brunswick, New Jersey

CRC Press
Taylor & Francis Group
Boca Raton London New York

CRC Press is an imprint of the
Taylor & Francis Group, an **informa** business

First published 1990 by Marcel Dekker, Inc.

Published 2019 by CRC Press
Taylor & Francis Group
6000 Broken Sound Parkway NW, Suite 300
Boca Raton, FL 33487-2742

© 1990 by Taylor & Francis Group, LLC
CRC Press is an imprint of Taylor & Francis Group, an Informa business

First issued in paperback 2019

No claim to original U.S. Government works

ISBN 13: 978-0-367-45082-3 (pbk)
ISBN 13: 978-0-8247-8274-0 (hbk)

Visit the Taylor & Francis Web site at
http://www.taylorandfrancis.com

and the CRC Press Web site at
http://www.crcpress.com

Library of Congress Cataloging-in-Publication Data

Sensors in Bioprocess Control / edited by John V. Twork, Alexander
  M. Yacynych.
       p.   cm. -- (Bioprocess technology; v. 6)
    Includes bibliographical references.
    ISBN 0-8247-8274-7 (alk. paper)
    1. Bioreactors. 2. Electrochemical sensors. I. Twork, John V.
      II. Yacynych, Alexander M. III. Series.
TP 248.25.B 55S 46   1990
660'.6--dc20   90-3103
                                                        CIP

Series Introduction

The revolutionary developments in recombinant DNA and hybridoma technologies that began in the mid-1970s have helped to spawn several hundred new business enterprises. Not all these companies are aimed at producing gene products or cell products, as such. Many are supportive in nature: that is, they provide contract research, processing equipment, and various other services in support of companies that actually produce cell products. With time, some small companies will probably drop out or be absorbed by larger, more established firms. Others will mature and manufacture their own product lines. As this evolution takes place, an explosive synergism among the various industries and the universities will result in the conversion of laboratory science into industrial processing. Such a movement, necessarily profit driven, will result in many benefits to humanity.

New bioprocessing techniques will be developed and more conventional ones will be revised because of the influence of the new biotechnology. As bioprocess technology evolves, there will be a need to provide substantive documentation of the developments for those who follow the field. It is expected that the technologies will continue to develop rapidly, just as the life sciences have developed rapidly over the past 10—15 years. No single book could cover all of these developments adequately. Indeed, any single book will be in need of replacement or revision every few years. Therefore, our continuing series in this rapidly moving field will document the growth of bioprocess technology as it happens.

The numerous cell products already in the marketplace, and the others expected to arrive, in most cases come from three types of bioreactors: (a) classical fermentation; (b) cell culture technology; and (c) enzyme bioreactors. Common to the production of all cell products or cell product analogs will be bioprocess control, downstream processing (recovery and purification), and bioproduct finishing and formulation. These major branches of bioprocess technology will be represented by cornerstone books, even though they may not appear first. Other subbranches will appear, and over time, the bioprocess technology "tree" will take shape and continue growing by natural selection.

W. Courtney McGregor

Preface

Since the first DNA fragments were successfully inserted into the
bacterium _Escherichia coli_ in 1973, the pace of discovery in molecular biology has been accelerating. Laboratory-based genetic engineering has produced an almost daily series of novel and exciting
developments and has redefined the role of microbes in agriculture,
pharmaceuticals, chemicals, energy, and waste treatment. Bring-
ing this advancing technology to the marketplace, however, requires important engineering advances to facilitate the successful
transition from the laboratory to production. This transition in
turn can only be realized by first gaining an understanding of the
conditions most favored for optimal microbial production and then
successfully duplicating and controlling those conditions on a large
scale.

Key elements in the optimization of biological processes include
the measurements of the growth rate, substrate uptake and production. Recent advances in both process analytical chemistry and sensor technology certainly promise to put into the hands of the biochemical engineer the tools necessary to monitor directly these important process variables.

It is our intention that this volume will not only present the
reader with an overview of current chemical sensor technology, but
also outline a practical framework relating industrial bioprocess monitoring to modern process control technology. Thus, the key element in many of the chapters describing on-line analytical methods
is their use (or potential use) in closed-loop process control; the
chapters introducing control technology and applications incorpo-
rate the coupling of chemical composition measurements. This effort
at interconnecting measurement science and process control technology is an attempt to introduce biologists and chemists to important
concerns in bioprocess control and to present control systems engineers with an overview of analytical methods capable of providing
a real-time measurement window on bioprocess chemistry.

The specific chapters dealing with chromatography, spectroscopy, electrochemistry, thermal analysis, and flow injection are not
intended to detail operational theory, which can be found in other

texts. Similarly, the chapters on conventional and advanced multivariable control technology focus on bioprocess applications. For extensive background in control theory, the reader is referred to other sources. In both cases, the effort has been made to present technologies that show considerable near-term promise. Beyond the specific treatment of bioprocess measurement and control, a number of related topics are addressed because of their importance to the overall success of implementing closed-loop control in the bioprocess industry. The chapters on justification and sampling should be of particular interest to those involved in the initial development and planning of projects concerned with improved process productivity or product quality.

It is our hope that, by addressing the wide range of issues involved in bioprocess control, we have underscored the importance of collaborative effort among diverse scientific and technical disciplines directed at the successful commercialization of biotechnology. It is also our hope that the technical content of this volume will prove a resource for those actively engaged in the effort to improve and optimize bioprocesses.

John V. Twork
Alexander M. Yacynych

Contents

Contributors

Bruce F. Bishop   Biological Sciences Department, Monsanto Corporation, St. Louis, Missouri

Steven D. Brown   Department of Chemistry and Biochemistry, University of Delaware, Newark, Delaware

René Bucher   Ingold Messtechnik AG, Urdorf, Switzerland

Hans W. Buehler   Ingold Messtechnik AG, Urdorf, Switzerland

Marvin Charles   Department of Chemical Engineering, Whitaker Laboratory, Lehigh University, Bethlehem, Pennsylvania

Kenneth J. Clevett   Clevett Associates   Inc., Watchung, New Jersey

Bruce Jon Compton   Fermentation Development Laboratories, Bristol-Myers Squibb Co., Syracuse, New York

George L. Eitel, Jr.* Project Management Department, Stone and Webster Engineering Corporation, Denver, Colorado

Robert J. Geise   Department of Chemistry, Rutgers, The State University of New Jersey, New Brunswick, New Jersey

Neil D. Jespersen   Chemistry Department, St. John's University, Jamaica, New York

Stephen J. Lorbert   Pilot Plant Engineering Support, Animal Sciences Division, Monsanto Agricultural Company, St. Louis, Missouri

---

*Present affiliation: Engineering Department, Lear Siegler Measurement Controls Corporation, Englewood, Colorado

John H. T. Luong   Biotechnology Research Institute, National Research Council of Canada, Montreal, Quebec, Canada

Ashok Mulchandani   Department of Biochemical Engineering, Biotechnology Research Institute, National Research Council of Canada, Montreal, Quebec, Canada

Arthur L. Reed   Control Equipment Corporation, Lowell, Massachusetts

Michael T. Reilly   Engineering Technology Laboratory, E. I. du Pont de Nemours & Company, Wilmington, Delaware

Otto S. Wolfbeis   Analytical Division, Institute of Organic Chemistry, Karl Franzens University, Graz, Austria

Alexander M. Yacynych   Department of Chemistry, Rutgers, The State University of New Jersey, New Brunswick, New Jersey

Sensors in
Bioprocess Control

1

The Needs for Sensors in Bacterial and Yeast Fermentations

BRUCE F. BISHOP *Monsanto Corporation, St. Louis, Missouri*

STEPHEN J. LORBERT *Monsanto Agricultural Company, St. Louis, Missouri*

I. INTRODUCTION

Recent advances in the field of biotechnology have greatly increased the need for new on-line sensor technology development. Recombinant DNA technology has resulted in the development of complex systems for the expression of heterologous proteins in bacteria, yeast, and mammalian cell culture. Frequently, the cell inserts the protein (protein products) into insoluble "refractile bodies," or the protein might be so unstable intracellularly that it is degraded by the cell as quickly as it is produced. Ideally, the fermentation is first carried to a high cell density and only then induced by either chemical or physical means to begin production of the foreign protein. In this manner, deleterious effects of the plasmid on the cell during growth can be minimized, and any toxic effects of the foreign protein on the cell will also be minimized.

Typically, these fermentations are composed of two distinct phases: a growth phase and an induction phase. Because of their dynamic nature, the fermentations (with growth rates and rapid doubling times approaching 30 min), new sensor technology must be developed to assist in monitoring and control. The ability to couple sensors to computers will allow sophisticated closed-loop control strategies to be developed which can utilize real-time data to automate and improve the process. The driving force toward higher levels of control will hopefully result in many process improvements, including higher final cell densities, an increase in the amount of product produced per unit of biomass, and sustaining consistent levels of increased productivity. Other benefits of improved control capability might include optimizing use of raw materials and improving product quality, which could dramatically decrease downstream purification costs.

A. Sensors for Cell Density Determination

The ability to monitor biomass concentration in a reactor is an important consideration in process development. In many recombinant
systems the process improvement objective is to maintain a consistent level of product expression per unit of biomass and to maintain
that level of expression at high cell density, thus increasing product yield. A second objective is to increase the level of product
expression per unit of biomass, again maintaining this level of expression at high cell density. In fed-batch fermentations, the critical nutrient(s) is(are) fed into the fermenter at, or near, the utilization rate of the organism. In this manner much higher final cell
densities can be achieved than in conventional batch processes.
The development of sensors to measure biomass concentrations in
the fermenter would prove invaluable in the development of strin-
gently controlled nutrient feed strategies.   These sensors will be
discussed in more detail in Section II. In fermentations of recombinant organisms, many process event markings can be a function of
biomass concentration. For example, as the biomass reaches a critical point for induction, it could signal addition of inducer to initiate expression of the desired protein. The possibility of morpholog-
ical changes in the cells upon induction must be considered in the
development of process event markings. In one study with
Escherichia coli, investigators found that induction of a foreign gene
(i.e., an analogue of human alpha interferon) caused a breakdown
in the correlation between culture turbidity and dry cell weight [1].
The breakdown appeared to be caused by the dramatic increase in
cell size as it produced high levels of the recombinant protein. Pri-
or to induction of alpha interferon there was a very good correlation
between culture turbidity and dry cell weight. If these phenomena
prove to be reproducible from batch to batch, computer control algorithms could be designed to use biomass data for process control
during the induction phase of the fermentation. Further, as the
fermentation becomes characterized and reproducible, deviations from
established growth profiles may also serve as indicators of contamination.
    Due to the absence of any method to determine biomass concen-
tration on-line, many off-line procedures are used to measure wet
cell weight, dry cell weight, viable cell count, and optical density
[2—4]. These methods can yield different results, depending on
the method used and on operator technique. Samples must be removed from the fermentor for assay off-line. Because of the time
required to generate assay data off-line and because of the dynamic
nature of these fermentations, matching control actions corresponding to critical time points in the fermentation will be very difficult,
if not impossible to achieve. Instruments have been designed to

automate sampling and measurement of culture turbidity. These systems involve either continuous or semicontinuous removal of fermentor broth to a dilution chamber where the culture is diluted to within the linear range of the spectrophotometer (usually 0.1—0.6 optical density units measured at 550 nm). The diluted sample is then automatically injected into the spectrophotometer for measurement. Major problems are encountered with fouling of the instrument tubing and orifices, bubble interference with the measurement, and the sterility of sampling, which can result in culture contamination.

A variety of novel technologies are being investigated for use in biomass monitoring. Laser technology and acoustic characteristics of biomass are being used to develop noninvasive sensors which would circumvent the major sensor disadvantages of sterilization/ contamination and on-line calibration. A review by Clark et al. [5] discusses these and other physical and biochemical principles on which new sensors might be based. The authors highlight techniques utilizing dielectric potential of microorganism membranes, infrared spectra of cultures, antibody-specific immunosensors and polarization of chiral structures such as DNA within the cell.

Probably one of the most promising areas of biomass sensor development is in the area of spectrofluorometry of cellular cofactors, such as reduced nicotinamide adenine dinucleotide (NADH). NADH is oxidized to NAD+ during oxidative phosphorylation for the production of energy-rich ATP to drive other metabolic pathways [6]. When irradiated with light at 340 nm, NADH fluoresces at 460 nm while oxidized NAD+ does not fluoresce. Duysens and Amesz [7] showed that culture fluorescence could be changed by altering the metabolic state of baker's yeast. Fluorescence profiles have been shown to be indicative of metabolic activity in yeasts and bacteria in a variety of fermentation systems [8—10]. The relationship between culture fluorescence and biomass concentration has been shown to have very good correlation [11], particularly in logarithmically growing cultures at relatively low cell density. Fluorescence technology has resulted in the development of commercially available fluorometers for biomass monitoring and are presently being evaluated in high cell density fermentations in industrial processes.

B. Sensors for Fermentation Medium Analysis

On-line compositional analysis of fermentation broth will allow the fermentation technologist to obtain critical information needed for process optimization. Sensors designed to monitor substrate concentrations could be used to calculate substrate utilization rates. Matched feeding of substrate to the organism's utilization rate would maximize growth and product formation. The need for stringent

control of residual glucose concentrations in the broth have been
shown to be critical for high product yields in recombinant strains
of E. coli [12] and Saccharomyces cerevisiae [13,14]. In addition,
the production of deleterious metabolic by-products such as organic
acids in bacteria and ethanol in yeast could be minimized with sen-
sor-based control.

Electrochemical Sensors

The lack of on-line sensors to detect sugars and other metabolites has led to the development of novel electrochemical sensors for
use in fermentation. These sensors contain immobilized enzymes on
an electrode to perform the analysis. Immobilized microorganisms
are also being considered for use in the development of electrochemical sensors for monitoring fermentation broths [15–17]. In a review by Twork and Yacynych [18], the authors list several advantages of electrochemical sensors: (a) selectivity is high, even in
turbid solutions; (b) the sensors can function over wide ranges of
concentrations of the chemical; (c) they can be used to monitor a
wide range of medium components, including organic acids and ethanol; and (d) electrochemical sensors are relatively cheap and easy
to operate. Electrochemical sensors have several disadvantages that
prevent their use in situ. They are not steam sterilizable and must
be incorporated into automated off-line analysis methods, resulting
in long lag times for assay. Contamination risks and plumbing
fouling are also potential problems in these systems. Enzyme analysis of certain components requires assay conditions outside of
physiological limits, e.g., ammonium ion control with an ammonia
electrode at pH 11.0–11.5 [19]. These analyses must be performed
outside the fermenter. On-line calibration of probes and stability
of the enzymes or microbes present additional problems. A number
of investigators have developed automated analysis systems utilizing
enzyme sensors [20–22]. Oguztoreli et al. [23] have developed a
mathematical model for optimizing substrate concentrations in fermentation broths with a separated sensor. Attempts are being made
to develop novel technologies to minimize the problems associated
with off-line automation. These approaches have been discussed
[18], and include the use of selective, sterilizable membranes for
providing assay material to the analyzer, chemically sterilized sensors, and, possibly, developing more stable enzymes.

High-Performance Liquid Chromatography (HPLC)

HPLC is a well-known, routine technique for the separation,
identification, and quantitation of chemical and biochemical compounds. A method has been developed to interface an HPLC with a

fermenter for continuous on-line monitoring of fermentation broths
[24]. A cross-flow membrane device is used to provide cell-free
broth to the HPLC for analysis. Depending on the chromatography
column configuration used, this "sensor" can be used to monitor
levels of sugars, alcohols, organic acids, or extracellular protein
concentrations. By monitoring protein concentrations in the broth,
an on-line HPLC could be used for event markings such as process
end-point determinations. An on-line HPLC system is commercially
available for industrial applications at costs approximately equal to
conventional HPLC systems. The added advantage of this system
is that it can also be used off-line in a manual injection mode. Gas
chromatography has also been used for monitoring volatile components [25] and for soluble metabolites in the broth [26]. Disadvantages of both HPLC and gas chromatography are that the connection of the system to the fermenter can be a source of contamination and the time between analyses can be several minutes, which
may be insufficient for control purposes. Adsorption of medium
components to the tubing and variability in sample volume automatically injected are additional sources for error.

Mass Spectroscopy

Mass spectrometers are used routinely in the fermentation industry to measure gas concentrations in the fermenter headspace.
Off-gas data is a dynamic measurement of metabolic activity of the
cells. Data are used in mass-balancing equations, such as carbon
dioxide evolution rate, oxygen uptake rate, respiratory quotient,
and oxygen transfer coefficient (K_{la}) [27]. Units are interfaced
with a computer and can be multiplexed to a number of fermenters.
Modification of a mass spectrometer to monitor dissolved volatile
gases has also been described [28]. A steam sterilizable membrane
probe is coupled to the mass spectrometer to provide a sample for
analysis. Mass spectrometers, however, cannot measure nonvolatile
compounds. Another disadvantage is that multiplexing capability is
not commercially available and can be expensive. Automatic control
of sequential sampling must also be custom designed for in-house
use. Mass spectrometers require significant capital investment, although less expensive units are becoming available.

II. SPECIFIC NEEDS FOR NEW SENSOR
   TECHNOLOGY

Current needs for new sensor technology are based on an ongoing
desire to increase our ability to monitor and ultimately control those
parameters which influence the productivity of the fermentation.

Previous topics presented in this chapter have provided insight into the major directions being taken for new sensor development. The technology emerging in this area is having a profound effect on our ability to monitor important environmental and physiological parameters. Our ability to collect a broader spectrum of information using less laborious means of analysis has been a real benefit in increasing our understanding of fermentation process phenomena. The focus for applying new sensor technology has initially been in the laboratory. As processes are moved from the benchtop and into the pilot plant, or into commercial production, the real question becomes, "How can new sensor technology be used to provide more effective control strategies for increasing fermenter productivities?" In this section, we hope to provide some insight into the needs and potential benefits of providing improved sensor technology for enhanced process control.

In a recent article on sensor development, the following statement was made by a researcher assessing the role of sensors in the development process, ". . . you scale up—to 10, 100, 1000 liters. You put the organisms into the tanks and pray. Today we can monitor pH, temperature, dissolved oxygen, and maybe something else if we're lucky. You try to get things in balance with only a few parameters. We have just scratched the surface of on-line monitoring" [29]. These statements are perhaps an overly conservative estimate of where the technology is currently, but does reflect a widely held view that new sensors have not yet made a major impact on our ability to implement more advanced control strategies for fermentation processes. From the information presented earlier in this chapter, there is no doubt that a wide variety of new monitoring instruments are becoming available. The main problem is that very few of these probes can be used on-line for process control. The monitoring information is not obtained easily and in a timely enough manner to be able to direct the course of the fermentation.

These limitations are especially evident in conjunction with the new generation of fermentation-based products evolving from biotechnology. The products emerging from this science are often based on fermentation processes involving yeast or bacteria. Typically, these organisms are fast growing with doubling times as short as 30 min. As a result, these fermentation processes have increasingly complex requirements for monitoring and control. Zabriskie and Arcuri [30] have reviewed the factors influencing the productivity of fermentations employing recombinant microorganisms. The authors state that initial solutions to productivity issues have been primarily based on genetics. However, in the long term, these approaches are likely to see diminishing returns. Ultimately, optimization will depend on a better understanding of the relationships between the microbial environment and product production. Optimization of environmental parameters beyond pH, temperature, and dissolved oxygen will require more advanced approaches for process control.

One of the initial areas of focus for process optimization has been the development of improved control strategies for carbon source feeding. It is well known that fermentation feeding strategies which promote high or uncontrolled levels of residual glucose can cause ethanol production in yeast, even under aerobic conditions [31—33]. A similar phenomenon has been documented in bacteria, except that organic acids are produced rather than ethanol [34]. Hollywood and Doelle have reported that this shift toward aerobic fermentation can result in acetate formation at glucose concentrations as low as 2 g/L [34—37]. Hence, tight control of the residual glucose concentration is a requirement if by-product production is to be avoided. Whether the system is yeast or bacterial, these by-products can lower growth rates, limit cell densities, and, in fermentations of recombinant organisms, limit the amount of product which can be produced during the course of the batch.

A number of approaches have been published for maximizing cell densities in yeast fermentations by minimizing the amount of ethanol produced during the course of the batch. Aiba, Nagai and Nishiywa, using respiratory quotient (RQ), the molar ratio of carbon dioxide produced to oxygen consumed, as a control parameter, developed an on/off control system for carbon source feeding [38]. The RQ value was maintained between 1.0 and 1.1 in order to minimize residual ethanol levels in the broth. Cooney et al. [39—40] have written a series of classic papers describing their use of an on-line computer based material balancing routine for estimating cell densities in yeast fermentations. The routine was further modified to allow for the implementation of an anticipatory control strategy for molasses feeding [41]. The approach worked well in the laboratory with final dry weights ranging from 50 to 60 g/L. In a more recent publication, Dairaku et al. presented a carbon source feeding strategy for yeast which is based on keeping a constant ethanol concentration during the fermentation [42]. The actual ethanol concentration in the broth was measured with a tubing method. The value of ethanol concentration was sent to a proportional, integral, derivative (PID) controller which regulated the glucose flow into the fermenter. The ethanol concentration could be controlled within 10 ppm by adjusting the glucose feed rate to maintain the net ethanol production rate at zero. Applying a single variable feedback control loop has potential advantages in terms of simplicity and maintenance compared with the more complex approaches referenced earlier.

As elegant as the foregoing strategies appear to be, they are finding limited application in the baker's yeast industry. In a re-

cent publication, Chen and Chiger state, "In practice, baker's
yeast is not propagated with strictly exponential substrate additions,
nor is it propagated at the highest specific growth rate compatible
with maximal substrate yield. Rather, baker's yeast propagation is
conducted to produce specific properties in the final yeast product"
[43]. The commercial baker's yeast fermentation process attempts
to maximize those characteristics in the yeast which give the greatest utility to the end product. Parameters such as leavening activity, appearance, and storage stability are not necessarily optimized under the same conditions that result in maximum biomass
production. This is also true of recombinant yeast and bacterial
fermentation systems where the conditions for product production
are not necessarily the same as those which maximize the production
of biomass. For example, in recombinant bacterial systems, a temperature shift can be employed for some types of vector systems to
increase the number of plasmid copies inside the cells. By increasing the average copy number, it is possible to amplify the expression of the recombinant protein [30,44]. Typically, the bacteria
are initially grown at a temperature that is optimal for growth.
When an appropriate cell density is reached in the fermentation, the
temperature is raised to cause an increase in the plasmid copy number. Ideally, this results in the expression of large amounts of the
recombinant protein in conjunction with a relatively low-level con-
stitutive promoter. In a sense, two fermentations must be optimized: one for growth and one for product expression. A good
carbon source feeding strategy is a primary factor influencing the
productivity of both phases of the fermentation. Feeding strategies
which minimize ethanol and organic acid production are likely to
have immediate benefits in yeast and bacterial fermentation processes
that are oriented toward the production of recombinant proteins.
From that viewpoint, the control strategies described earlier for
baker's yeast may be readily applicable to yeast fermentation processes that are directed toward the production of recombinant proteins. At this point, however, there is no published information describing the use of such strategies for yeast.

Carbon source feeding as it relates to the production of recombinant proteins in bacterial systems has been discussed by few authors [30,45,46]. In general, the prevailing opinion is that controlled carbon source feeding is an important factor in minimizing
by-product production and maximizing the expression of the recombinant protein of interest. The most effective strategies appear to
be those which try to control feed rates based on measurements of
residual glucose levels in the fermenter. It does not appear that
there is yet a sterilizable probe which can be inserted into the fermenter for on-line monitoring of carbon source. Rather, most of the
development work is being directed toward the development of asep-

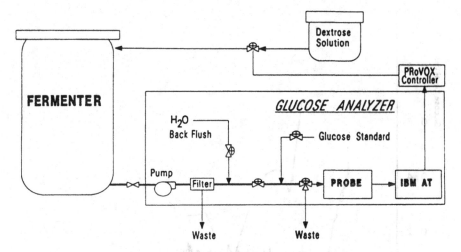

Figure 1   Glucose analyzer/control system (From Ref. 46 with permission).

tic sampling systems which can deliver an acceptable sample to an off-line instrument for analysis.   A number of recent papers provide a perspective on the progress being made in implementing this type of an approach [20–22].   Figure 1 provides a schematic of the system developed by Aynardi et al. [46].

Figure 2 compares residual glucose concentrations in fed batch E. coli fermentations using the system depicted in Figure 1.   Also depicted is the glucose profile from a batch using a simple programmed feeding strategy where set-point adjustments were controlled by the operator.   The potential benefits for limiting organic acid production by using a more controlled approach to glucose feeding are obvious from Fig. 3.   As is evident from the references already cited, there has been significant progress made in developing instrument/control systems which can improve carbon source feeding.   Additional progress will be necessary before these systems can be implemented in commercial fermentation processes.

Based on the preceding discussion, it is evident that new instrumentation like that being developed for carbon source monitoring will enhance our ability to better control basic environmental conditions during the fermentation.   Microbial growth and the expression of products directly depend on the degree to which we can manipulate and control these external parameters.   Optimization of fermentation conditions will lead to greater productivities, regardless of the scale at which the fermentation is being run.   Optimization of fermentation conditions is usually performed in the process development phase, with the overall strategy being developed in the labo-

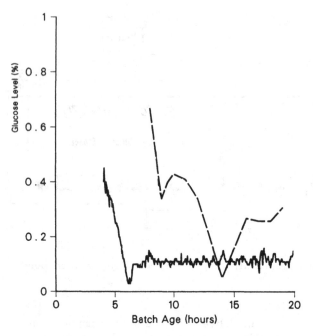

Figure 2 Fermentation glucose concentration profiles (From Ref. 46 with permission).

ratory or pilot plant. Once the process is transferred to a commercial setting, it is run in a routine and, hopefully, reproducible manner to generate sufficient quantities of the product of interest. Reproducibility is a key factor in making a potential fermentation strategy viable for commercialization.

Reproducibility in yeast and bacterial fermentations is often linked to event timing and the degree to which the operator can make a specific operation occur at a given time during the fermentation. Historically, the operator will less likely take action based on time and more likely perform an operation based on cell density. This is certainly true of the newer generation of fermentation processes for the production of recombinant protein products. As mentioned previously, it is common in these fermentations to grow the culture in a cell division mode until a target concentration is reached in the fermenter. The cells are then shifted from a cell division mode into a product production phase [49]. Khosrovi and Gray present actual growth curve data for E. coli producing interleukin-2 (IL-2) [49]. The cells are initially propagated at a temperature at which IL-2 synthesis is repressed. A shift of temperature to 42°C

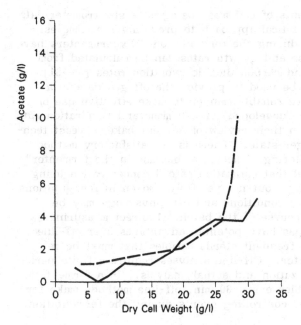

Figure 3 Fermentation by-product production (From Ref. 46 with permission).

induces rapid synthesis of the product. This shift in operating temperature is made once a sufficient cell density is reached in the fermentation. Both the growth and product expression phases of the fermentation are monitored by using optical density.

It has been pointed out earlier that optical density can be used as a quick off-line technique for assessing biomass concentration. Traditionally, this method has worked well and appears especially satisfactory in those processes where cell densities are not very high. Higher cell concentrations require greater dilution of the broth sample, potentially reducing the overall accuracy of the analysis. On-line monitoring of optical density (OD) is possible only at lower cell concentrations. These limitations are being addressed as a result of new technology which uses a laser light source for OD measurements. (See previous discussions in this chapter on optical density sensors.) Additional difficulties can be encountered in using optical density to measure cell mass in fermentations where the bacteria are producing large amounts of a precipitated intracellular product [1]. These problems can be obviated by having a good correlation of dry cell weight and optical density for the given fermentation.

Indirect measurements of cell mass using mass spectrometer data may provide a more practical approach to providing real-time estimates of cell densities during the fermentation. Experimenters have established that biomass and growth rates can be calculated from oxygen uptake rates and carbon dioxide evolution rates [50—51]. A mass spectrometer can be used to provide the off-gas data in a timely and perhaps more reliable manner to make effective use of correlations that can be developed for cell mass and respiration rate parameters [52—53]. In their review of current baker's yeast technology, Chen and Chiger state, "There is no satisfactory method for direct on-line monitoring of microbial biomass in the fermenter" [43]. The authors feel that estimation of cell concentration using off-gas data is promising, but not yet fully proven in fermentations where the environmental conditions and cell physiology may be changing through the course of the batch. Indirect measuring methods based on off-gas have potential advantages over off-line methods which require frequent sterile samples that must be removed from the fermenter. Off-line analyses require a finite period for sample valve sterilization and actual analysis. In a fermentation where the culture doubles every 30 min, off-line methods make for a less reliable basis for controlling events during the fermentation.

III. ECONOMIC IMPACT OF NEW SENSOR
      TECHNOLOGY

When the fermentation can be run reproducibly and in an optimized manner, there are obvious economic benefits that can be realized for commercial processes. By ensuring that the process consistently maintains productivities in a targeted range, a good process control strategy can lead to lower capital expenditures for a new production process. Batch sizes can be minimized if the variation in product expression is also minimized. By optimizing batch volumes, the process engineer can provide a more cost effective design for the fermenter and associated media vessels. Lower production volumes lead to lower installed horsepower costs for agitation systems. Associated utility systems can also be downsized when process variability is minimized through better process control. Of equal benefit is the potential impact on the sizing of downstream purification facilities. Tighter estimation of batch sizes and productivities in fermentation will result in a more cost-effective design for product recovery operations. References are available which discuss economics and cost estimation for fermentation-based processes [54—56]. To a lesser degree, these references also discuss the interactions between fermentation and purification, and the potential cost impact of one on the other.

A less variable fermentation process is also likely to result in improved operating costs for both fermentation and product purification. Maximizing process reproducibility and productivities in fermentation will result in optimum raw materials used in that area of the process. Likewise, being able to deliver a fermentation product of consistent quantity and quality can have a major influence on raw materials used in purification. Raw materials costs are always a major factor influencing overall manufacturing costs for a process. This is especially true of the newer generation of products evolving from biotechnology. Fermentation media for this new generation of products often consists of a mixture of defined ingredients that make for a more controlled, but also more costly, fermentation. Product purification economics are also more sensitive to the costly solvents and chromatography resins that seem generic to the purification processes which have been developed for these products. Datar has estimated that the ratio of purification to fermentation costs for these processes is about 2.0 [55]. This ratio is even higher for human pharmaceutical products. Hence, improved process control can lead to improved productivities and product quality in fermentation. These benefits can potentially have even greater returns in terms of the economic impact on the downstream portions of the process.

IV. CONCLUSIONS

Overall, improved sensor technology can provide the basis for new control strategies which can increase our ability to run less variable, more productive fermentations. New sensor technology is developing rapidly, as evidenced by the increasing number of new probes which have become available for measuring a wide variety of parameters. The challenge at this time is developing sensors which can be readily implemented for more effective process control. Improvements in our ability to control the fermentation will result in more consistent and more productive fermentation strategies. These improvements will have immediate economic benefits that can lead to more cost-effective production processes.

REFERENCES

1. Fieschko, J., Ritch, T., Bengston, D., Fenton, D., and Mann, M. *Biotech. Prog.*, 1, 205 (1985).
2. Gavin, J. J. *Appl. Microbiol.*, 5, 235 (1957).
3. Gerhardt, P., Murray, R. G. E., Costilow, R. N., Nester, E. W., Wood, W. A., Krieg, N. R., and Phillips, G. B. *Man-*

ual of Methods for General Bacteriology, American Society for
Microbiology, Washington, D.C. (1981), pp. 182—197.

4. Pirt, J. S. Principles of Microbe and Cell Cultivation, Black-
    well, London (1975), pp. 15—21.

5. Clark, D. J., Blake-Coleman, B. C., Carr, R. T. G., Calder,
    M. R., and Atkison, T. Trends Biotech. 4, 173 (1986).

6. Lehninger, A. L. Biochemistry, 2d ed., Worth, New York
    (1975).

7. Duysens, L. N. M. and Amesz, J. Biochem. Biophys. Acta.
    1957; 24:19—26.

8. Armiger, W. B., Forro, J. F., Montalvo, L. M., Lee, J. F.,
    and Zabriskie, D. W. Chem. Eng. Commun., 45, 197 (1986).

9. Luong, J. H. T. and Carrier, D. J. Appl. Microbiol. Bio-
    technol., 24, 65 (1986).

10. Harrison, D. E. F. and Chance, B. Appl. Microbiol., 19,
    446 (1970).

11. Zabriskie, D. W. and Humphrey, A. E. Appl. Environ. Mi-
    crobiol., 35, 336 (1978).

12. Moks, T., Abrahmsen, L., Osterlof, B., Josephson, S.,
    Ostling, M., Enfors, S. O., Persson, I., Nilsson, B., and
    Uhlen, M. Biotechnology., 5, 379 (1987).

13. Fieschko, J. C., Egan, K. M., Ritch, T., Koski, R. A.,
    Jones, M., and Bitter, G. A. Biotechnol. Bioeng., 29, 1113
    (1987).

14. Carty, C. E., Kovach, F. X., McAleer, W. J., and Maigetter,
    R. Z. J. Indust. Microbiol., 2, 117 (1987).

15. Karube, I. and Suzuki, S. Ann. Rep. Ferm. Proc., 6, 203
    (1983).

16. Hikuma, M., Kubo, T. and Yasuda, T. Biotechnol. Bioeng.,
    21, 1845 (1979).

17. Mullen, W. H. and Vadgama, P. M. J. Appl. Bact., 61, 181
    (1986).

18. Twork, J. V. and Yacynych, A. M. Biotechnol. Prog., 2, 67
    (1986).

19. Thompson, B. G., Kole, M., and Gerson, D. F. Biotechnol.
    Bioeng., 27, 818 (1985).

20. Kernevez, J. P., Konate, L., and Romette, U. T. C. Bio-
    technol. Bioeng., 25, 845 (1983).

21. Chotani, G. and Constantinides, A. Biotechnol. Bioeng., 24,
    2743 (1982).

22. Mandenius, C. F., Danielsson, B., and Mattiasson, B. Bio-
    technol. Letters, 3, 629 (1981).

23. Oguztoreli, M. N., Ozum, B., and Gerson, D. F. Biotechnol.
    Bioeng., 28, 952 (1986).

24. Dincer, A. K., Manohar, K., Skea, W., Ryan, M., and Kier-
    stead, T. Devel. Indust. Microbiol., 25, 603 (1984).

25. Comerbach, D. M. and Bu'Lock, J. D. Biotechnol. Bioeng., 25, 2503 (1983).
26. McLaughlin, J. K., Meyer, C. L., and Papoutsakis, E. T. Biotechnol. Bioeng., 27 1246 (1985).
27. Buckland, B., Brix, T., Fastert, H., Gbewonyo, K., Hunt, G., and Jain, D. Biotechnology, 3, 982 (1985).
28. Heinzle, E., Furukawa, K., Dunn, I. J., and Bourne, J. R. Biotechnology, 1, 181 (1983).
29. Van Brunt, J., Biotechnology, 5, 437 (1987).
30. Zabriskie, D. and Arcuri, E., Enzyme and Microb. Tech., 8, 705 (1986).
31. Crabtree, H., Biochem J, 23, 536 (1929).
32. Epps, H. and Gale, G. Biochem J., 36, 619 (1942).
33. Beck, C. and Von Meyenburg, H., J. Bact., 96, 479 (1968).
34. Doelle, H., Ewings, K., and Hollywood, N., in Advances in Biochemical Engineering and Biotechnology, No. 23 (1982), pp. 1–35.
35. Doelle, H., Hollywood, N., and Westwood, A., Microbios, 9, 221 (1974).
36. Doelle, H. and Hollywood, N., Microbios, 21, 47 (1978).
37. Hollywood, N. and Doelle, H., Microbios, 17, 23 (1977).
38. Aiba, S., Nagai, S., and Nishyawa, Biotechnology and Bioengineering, 17, 1001 (1976).
39. Cooney, C., Wang, H., and Wang, D., Biotechnology and Bioengineering, 19, 55 (1977).
40. Wang, H., Cooney, C., and Wang, D., Biotechnology and Bioengineering, 19, 68 (1977).
41. Wang, H., Cooney, C., and Wang, D., Biotechnology and Bioengineering, 23, 975 (1979).
42. Dairaku, K., Yamasaki, Y., and Kuki, K., Biotechnology and Bioengineering, 23, 2069 (1981).
43. Chen, S. and Chiger, M., in Comprehensive Biotechnology, (Moo-Young, ed.), Pergamon Press, New York (1985), pp. 430–461.
44. Nicaud, J., Mackman, N., and Holland, I., Escherichia coli, Journal of Biotechnology, 3, 175 (1985).
45. Battelle Corp. Publication, "A Gradient Feed Process for Obtaining High Cell Densities with Reduced Acid Production for Eschericia coli," Presented at the Annual Meeting of the American Society of Microbiology (1986).
46. Aynardi, L., Lorbert, S., Pike, J., Roettger, B., Striebel, J., and Ziha, J., "An Optimized Glucose Feeding Strategy for Fed Batch Cultures of E. coli," Presented at the Annual Meeting of the AICHE (1987).
47. Reilly, M., Charles, M., and Phillips, J., "Development and Implementation of a Feedback Glucose Controller for a Fed

Batch E. coli Fermentation," Presented at the Annual Meeting of the AICHE (1987).

48. Ghoul, M., Ronat, E., and Engasser, J., Biotechnology and Bioengineering, 28, 119 (1986).
49. Khosrovi, B. and Gray, P., in Comprehensive Biotechnology, Vol. 3 (Moo-Young, ed.), Pergamon Press, New York (1985), pp. 319−330.
50. Meyer, H., Kappeli, O., and Fiechter, A., Ann. Rev. Microbio., 39, 299 (1985).
51. Zabriskie, D. and Humphrey, A., AICHE J., 24, 138 (1978).
52. Armiger, W., in Comprehensive Biotechnology, Vol. 1 (Moo-Young, ed.), Pergamon Press, New York (1985), pp. 133−148.
53. Bull, D., Comprehensive Biotechnology, Vol. 1 (Moo-Young, ed.), Pergamon Press, New York (1985), pp. 149−163.
54. Stanbury, P. and Whitaker, A., Principles of Fermentation Technology, Pergamon Press, New York (1984).
55. Datar, R., Process Biochemistry, Feb., 19 (1986).
56. Kalk, J. and Langlykke, A., Manual of Industrial Microbiology and Biotechnology, (1985), pp. 363−384.

2
Sampling

GEORGE L. EITEL, JR.* *Stone and Webster Engineering Corporation,*
Denver, Colorado

I. INTRODUCTION

A. Development of Aseptic Sampling

From the earliest recorded history of man, samples of materials and
substances have always been taken to represent the bulk quantity
in trading transactions and direct sales.   Special sampling considerations have developed from simple procedures such as keeping
food samples cool to reduce the rate of spoilage and have evolved to
modern techniques of freeze-drying, drying, and antibiotic drug and
ultraviolet ray treatment for vastly improved sample storage and
preservation. Microorganisms such as yeasts, bacteria, and molds
need to be controlled, inactivated, or removed to meet today's regulatory requirements for food and drug products.

   In the field of medicine and human health the control of these
microorganisms has been a serious problem.   In the mid-nineteenth
century Louis Pasteur discovered that bacteria caused disease, and
the British surgeon Joseph Lister pioneered the scientific application
of antiseptics in surgery, based on the earlier discoveries of Pasteur.   To prevent infection, Lister developed antiseptics such as
carbolic acid to disinfect operating rooms and patients' wounds to
prevent bacterial infections.   Since that time significant discoveries
have enabled scientists to develop aseptic techniques to further control microorganisms during medical treatment.

   The pharmaceutical industry manufacture of human drugs or
parenterals requires effective sterilization techniques using chemicals, saturated steam, dry heat, irradiation, and other methods for
eliminating unwanted microorganisms.   These sterilization methods
have been applied in the rapidly growing biological process industries.   The requirement for aseptic conditions in biological reactions,
aseptic media fills, and final parenteral drug packaging is now reg-

---

*Present affiliation: Lear Siegler Measurement Controls Corpora-
tion, Englewood, Colorado

ulated by agencies such as the Food and Drug Administration (FDA). The use of current good manufacturing practices (CGMPs) and good laboratory practices is designed to apply proper methods of sterilization for specific biological processes.

At the present time, the art of aseptic sampling in biotechnical facilities is expanding, with equipment manufacturers offering manual and automatic sampling devices for installation in the biological fermentation equipment, bioreactors, tanks, and associated equipment. Standard procedures for performing aseptic sampling is primarily the responsibility of the user of the equipment, although some guidelines are provided by the equipment manufacturer and engineer/designer of the plant.

However, a wide variation in methods for sampling these process equipment exists today, and more uniform methods and procedures are needed in the industry. The variety of processing and equipment requirements in this new industry complicates this goal as standardization efforts continue to grow. For example, the American Society for Testing and Materials (ASTM) has organized the E-48 Committee on Biotechnology during the past four years to address standard test methods and procedures for identification of microorganisms, terminology, and process control to containment, environmental, toxicity, and disposal issues. ASTM currently has a draft Standard Practice for Aseptic Sampling being developed for eventual use by the industry. Of the 5000 ASTM standards currently available, about 200 address sampling, but the E-48 committee is producing the first aseptic sampling standard practice for industrial application.

B. Challenges in Proper Sampling

Although the CGMPs vary from country to country regarding application to genetically engineered processes, it is concluded that the industry application of CGMP guidelines is growing steadily. One factor influencing this conformance is the increasing compliance enforcement by regulating agencies such as the FDA and USDA. As new processes are advanced from the laboratory to the commercial plant, it becomes important that process procedures and equipment be validated at all stages of development. As equipment scale-up progresses, the sampling methods also change as the reactor batch size increases from a few milliliters to over a thousand liters. This requires full coordination between the molecular biologist, genetic engineer, and plant engineer so that the process and its sampling methods are fully validated and the equipment qualified for repeatability and accuracy.

The primary goal for any sampling system, aseptic or not, is to obtain a representative sample that is properly handled so that the desired analytical test can be conducted. Thereby, an accurate analysis is obtained representing the bulk material quality at the specific time that the sample was obtained. The sampling procedure design should include the following analytical results, using the best available technology and safety standards:

1. Quality
2. Identity
3. Purity
4. Concentration
5. Repeatability of results

These results will then provide information for process control within a prescribed set of tolerences.

Since some genetically engineered products are toxic or their impact on the environment has yet to be determined, another challenge would be to properly design containment facilities for the sampling system to control emissions, purges, resterilization of effluent, sample handling, storage, and sample disposal after the desired analyses are complete.

Another challenge for accurate sample retrieval and handling is in the training and proficiency of the personnel involved in the sampling activities. Since sampling procedures should be written, updated periodically, and validated, it follows that initial personnel training should be followed by frequent observation and periodic refresher training. Aseptic sampling is also partly an art to achieve reproducible results. Successful operator techniques should be reviewed and included in the written sampling procedure.

Aseptic processing guidelines can be used if the sampling of the biosystem precludes the introduction of foreign microorganisms into either the bulk system or the sample. These guidelines are provided for drug products in Title 21 Code of Federal Regulations 1, parts 210 and 211. For biological products the drug guidelines can apply as well as parts 600 through 680. Sterility testing of biological products and their culture media used in such testing must conform to the requirements under section 610.12 when the product must comply with FDA regulations. Other suitable criteria from other regulatory agencies and professional organizations can be used as appropriate.

C. Sampling Categories for Biosystems

Two general types of sampling methods are normally used in biosystems and drug processes:

1. Indirect—the sample is removed from the bulk material and placed into an external container for transport to the analytical testing area.
2. Direct—the analytical test is conducted directly in the bulk material without a sample being removed. This is also called in situ sampling.

For either indirect or direct sampling techniques there are two broad categories of sterility involved which depend on the nature of the process and its specific sterility requirements. Aseptic conditions require adequate control over microorganisms, pressure, temperature, dust, humidity, and air quality. The two categories of acceptable exposure recognized by FDA [1] are

1. Critical areas—the most stringent criteria such as sterile room design requirements using high-efficiency particulate air filter (HEPA) filtered laminar flow air of high microbial quality (no more than one colony forming unit per 10 ft^3 for a class 100 rating).
2. Controlled areas—where unsterilized materials are exposed to the plant environment. Class 100,000 criteria are applicable with microbial activity of no more than 25 colony forming units per 10 ft^3.

D. Sources of Sampling Errors

Since many of the problems associated with some analytical methods have been attributed to the sampling method and handling problems, the following potential sampling errors are listed for consideration:

1. Adequate sample quantity must be provided.
2. Contamination of sample with extraneous material must be avoided.
3. A representative sample must be obtained from bulk material.
4. Uniform sample collection methods must be used.
5. Adequate written sampling procedures are needed for specific systems and organisms.
6. Portions of aseptic sampling system must be designed properly, such as multiple sample draw where resterilization of associated sample devices is not possible.
7. Adequate personnel training is essential.
8. Adequate employee hygiene and protective equipment are needed.
9. Proper sample handling is required from collection to the time sample is analyzed. For example, some samples should be fro-

zen or cooled upon collection to stop the reactions and pre-
serve the sample quality while waiting for the analytical testing to be completed.

10. Authorized procedure changes or revised application must include updating the sampling protocol.

11. Validate the system using microbiological challenges to simulate
worst-case production conditions. Microorganisms should be
selected to simulate the smallest microorganism that may occur
in production where sterile environment is needed.

12. Where aseptic sampling equipment is needed and this equipment
is removed for steam sterilization in autoclaves, it is important
that established loading patterns be followed repeatedly to
maintain the process validation.

13. Microbiological culture media used in environmental monitoring
should be capable of detecting molds and yeast as well as bacteria.

14. Identification of the organism in the sterility tests is needed at
least by its genus. Trends in system bioburden levels should
be detectable in monitoring the system sterility.

15. Sampling system design or operating problems should be reviewed to avoid dead zones of material which can generate bacteria or otherwise contaminate the next sample drawn through
the device.

16. Proper design is essential for sample bottles to withstand
sterilization and process operating conditions.

17. Other considerations as described in the following sections of
this chapter.

II. GENERAL CONSIDERATIONS

A. Safety

Sampling systems should be designed and operated, using the appropriate safety regulations for the process and materials being
used. Some of the safety considerations include

1. Following established worker safety precautions from OSHA,
state and local agencies, fire marshals, USDA, FDA, NIH, professional societies, ASTM, insurance company, and other resources pertinent to the biotechnical industry. These guidelines should include equipment design considerations for sampling device access by the operator, protective enclosures to
contain the sample material, protective clothing for the operators, material data sheets for reference and training to help
identify the potential hazards of the materials being handled,

fire protection system for the process, fire fighting and spill
containment considerations, emergency first aid procedures for
exposure to any hazardous materials and other pertinent considerations.

2. Using written, approved operating procedures for all sampling
    activities from sterilizing the sampling devices, drawing the
    sample, preparing the sample container, transporting the sample
    for analysis, storing the sample before and after analysis, and
    disposing the sample and container after use.

3. Providing initial operator training and refresher reviews on
    sampling procedures. Written changes and training updates
    should be initiated whenever process changes or equipment mod-
    ifications are made that change the sampling procedures.

4. Using appropriate environmental monitoring for sampling of tox-
    ic and hazardous materials.

B. Disposal Considerations

Special precautions and procedures should be developed for disposal
of biologically active waste, samples, effluent generated during
sterilization of sample points and sample containers when not re-
used, vented materials from sample collection activities, decontami-
nation of spillage, and handling of unintentional releases of biological materials to the environment. These procedures should be
written, qualified, and incorporated into operating and emergency
procedures. Appropriate operating personnel and others, such as
in-plant and community fire fighters, should receive training in
these methods on a timely basis. These procedures should have the
appropriate federal, state, and local reviews and approvals.

C. Sample is to Be Representative

Two general categories of biological systems are normally sampled
as described below:

1. <u>Homogeneous system</u>—composed of a single-phase material such
    as a solid, liquid, or vapor.

2. <u>Heterogeneous system</u>—contains significant mixtures of two or
    more phases. A slurry of solid materials in a liquid medium re-
    quires special design and operating considerations to ensure
    that a representative sample of the bulk material is obtained.
    This slurry also requires special consideration for removing the
    sample from the container for analysis. Depending on the ana-
    lytical tests to be performed on the sample, it may be possible
    to remove the solids by filtration as in the case of pH measure-
    ments. If the analytical test does not require a sterile sample,

then the prime consideration for the sample device design shifts
to protecting the biosystem bulk material from inadvertent microbial contamination by the sample system.

In either category it is critical that the sample obtained be representative of the bulk material. Normally, the average quality of the
entire bulk material is desired, which requires adequate mixing of
the contents of the container and withdrawal of a representative
sample without altering its quality and composition. Sometimes the
sampling objectives could require samples from potentially stratified
zones of the bioreactor to check for sedimentation or specific layers
of organisms, such as floating yeast colonies.

D. Sample Size

The size of the sample should be set by the minimum quantities required to satisfy the following criteria:

1. Minimum sample required by the analytical tests. This would include the primary test being conducted and any supporting tests.
2. Sample quantity needed to make up for losses in sample preparation. Losses due to filtration, distillation, and other preparation steps should be considered.
3. Losses due to flushing the sample device prior to catching the
   sample. This could also include sample holdup in the sampling
   system which cannot be returned to the bulk material.
4. Amount of sample remaining as a "heel" in the sample container
   after the specified test sample is removed. This quantity can
   be used for sample container wall wetting when appropriate to
   minimize materials sticking to these walls and introducing errors
   into the analysis, such as sticky mammalian cells adhering to
   the walls when the sample is being removed for a cell count.
5. Sample storage requirements and retained samples should be
   considered for future analysis. Referee sample requirements
   should be considered. A safety factor may be needed to protect for retesting possibilities.
6. Additional material requirements for sample shipments.
7. Time available to obtain the sample. If the reaction kinetics are
   fast, time may not be available to collect a larger sample, and
   it will have to be frozen for delayed analyses.

After all these requirements have been identified, the quantity
of sample to be removed must be reconciled with the volume of bulk
material available in the process, the cost of the material being
sampled, the risk of having an inadequate sample quantity which
cannot be supplemented later when the process conditions have

changed, the risk of upsetting the process equilibrium by removal
of too much sample, and the capacity of the sampling system to draw
the required sample quantity without changing the composition of
the sample (i.e., the need for multiple sample draws from different
sample taps or where a time delay in multiple draws could change
the quality of a biosystem composite sample).

E. Sterility Requirements

The degree of sterility must be established for each process and
material being sampled. This decision could be based on the sterility of the final product, corporate sterility objectives, sensitivity of
intermediate process steps to potential contamination and its effects
on the final product quality requirements, CGMP considerations,
equipment containment design, risk of the sampling system causing
contamination of the sample or its source, impacts of sample residue
remaining in the sampling device, availability of a satisfactory method to resterilize the sampling device after each use, and other related considerations.

The sampling devices described in this chapter generally reflect
aseptic sampling conditions. Relaxation of these criteria would be
evaluated for the specific bioprocess system being considered.

F. Utilities Considerations

Utilities associated with sterilization of the aseptic sampling devices
generally include

1. Sterile steam
2. Sterile water for injection (WFI)
3. Sterile hot air
4. Ethylene oxide

Sterilizing steam is commonly used to clean and sterilize bioprocess equipment and their sample devices. Steam quality should
be free of chemical contamination such as amines, which can foul
bioreactor surfaces and possibly affect microorganism growth. Normally, proper steam conditions for sterilization require saturated
steam at 15 psig and 121°C. It is desirable to have 100% saturated
steam, not superheated or supersaturated. Superheated steam does
not penetrate as well, and its effectiveness against certain microorganisms at 121°C is inadequate for sterilization without extended
sterilization time. Supersaturated steam can also extend the sterilization cycle and cause condensation in the equipment, which can en-
courage bacteria growth compared with a saturated or dry system.

Sterile water for injection can be used for appropriate wash cycles during the sterilization process. This water quality is often used to produce the sterile steam.

Hot sterile air is another source for sterilizing sample devices and equipment. However, this method requires additional time for the sterilization. USP recommends using hot air at 170°C for at least 2 hrs. Sample bottles can be sterilized using hot air or steam in autoclaves.

If containment chambers are used around sampling systems, sterile cool air may be used for environmental control. Since HEPA-filtered air requires significant investment, careful evaluation of its required use is recommended, as with the overall decision to sterilize the sampling system or not to sterilize.

Ethylene oxide sterilization is another alternative for sterilizing sampling devices that can be removed, disassembled, and placed in the autoclave using ethylene oxide. If the sampling device is part of a large commercial bioreactor system, it may be more practical to use sterilizing steam instead of ethylene oxide.

C. In-Line, On-Line, and Off-Line Testing

Generally, the decision to conduct in situ analytical tests centers on the ability of the instrument to withstand the process environment. The instrument should be checked for signal drift and recalibrated in place if necessary. Also it should withstand sterilization conditions. Normally, in-line testing is conventional for in situ measurements of temperature, pH determinations where nonfouling liquids are used, pressure, agitator speed and power, dissolved oxygen, foam, gas flow rates, level, dissolved carbon dioxide, redox, and intracellular NADH.

Off-line or analytical devices requiring sample removal for testing usually are difficult to sterilize in situ or are complex, bulky devices that require taking the sample to the instrument. Cooney and Humphrey [2] report a special class of measurements requiring an on-line sampling technique using a recirculation loop. Care must be taken in sampling system design to avoid errors of poor aeration, poor mixing, and time delays. Analyses included in this category are paramagnetic analyzers for carbon dioxide and ammonia, gas chromatographs, spectroscopy, flow microfluorometry, dialysis and continuous filtration, multiple internal reflection spectrometry (MIR), and chromatography.

The limitations of many analytical devices and methods are due to limited ability to collect a representative sample and to employ a sterilizable probe. Accordingly, significant challenges remain for the scientific community to solve problems related to sampling procedures and devices.

H. Liquid versus Vapor Samples

Most of this discussion has been centered on liquid-phase samples. Vapor-phase and aerosol sampling have similar sampling criteria, sterility objectives, and considerations to obtain a representative sample. However, the collection methods are somewhat different to accommodate the vapor phase of the sample. Almost any method is an improvement over gravity settling, where the solids or liquids in the vapor settle out naturally over a long period of time, with consequent poor reproducibility and nonuniform settling rates.

Other collection methods for aerosols include using inertial devices with impact plates to knock out the entrained particles, centrifugal devices using helical paths for separations, precipitators (electric or thermal), filters, and membrane filters.

I. Sample Preservation

Depending on the specific process design, microorganisms involved, sterility requirements, kinetics of the reactions, and other considerations, the sample removed from the biosystem may undergo several special treatments prior to conducting the analytical tests. If the time lag between sampling and testing is too long, it may be necessary to stop the reaction by altering the physical properties of the sample such as changing the pH, temperature (cooling or freezing the sample), filtering the active cells from the broth, or other specific changes to the sample which do not impact the test objectives.

Normally, a key consideration for preserving sample purity will be the use of a sterile sample device and container to avoid introducing contamination and foreign microorganisms which could alter the sample quality and invalidate future test results. Long-term storage of the sample would require using appropriate conditions for the specific process fluids involved to extend the shelf life of the sample.

III. TYPICAL STERILE SAMPLING DEVICE DESIGNS

A. Design Considerations

There are a significant number of design variables that influence the type of sample system for a specific biosystem. These variables can be categorized as shown below:

Sampler location:
   Sample withdrawal position
      method of connection
      fluid velocity profile
      containment considerations
Sampler and container design
   Materials of construction
   Process and sample variables
      temperature
      pressure
      slurry/two phases
      viscosity
   Sterilization options
   Purging considerations
   Venting considerations
   Containment
   Design options

Sampler Location Considerations

Probably the single most important consideration in obtaining a representative sample of the bulk fluid is to properly locate the sampling device on the equipment. Some basic methods for attaching the sample collective device properly and improperly are shown in Figures 1, 2, 3.

Figure 1 shows a typical turbulent flow profile inside a pipe and potential sample connection designs. As one would expect, the maximum fluid velocity should be in the center of the pipe as represented by V-1. This velocity decreases as the wall of the pipe is approached. Velocity point V-2 represents the average velocity of the fluid in the pipe. Both V-1 and V-2 are assumed to be in turbulent flow so that the fluid composition is uniformly mixed. Point V-3 is adjacent to the wall and is in laminar flow. The respective sample taps for these velocity profile points are X, Y, and Z. If the fluid in this pipe is a slurry which requires turbulent flow to maintain a uniform, representative mixture, samples obtained from points X and Y should be characteristic of the average fluid quality. However, samples from point Z probably would not be representative of the slurry, since the contained solids could drop out of this boundary layer at laminar flows. If this laminar layer composition changes significantly from the average of the bulk material, the composition at points X and Y may also be altered. This laminar flow characteristic exists in any process equipment where there is a fluid flow. Design of the sample tap must consider these potential slurry distribution problems. As the viscosity of the fluid increases, this boundary layer problem increases. If sticky cells or other

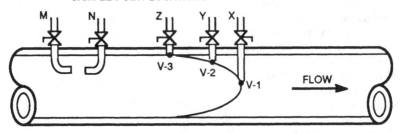

SAMPLE POINT LOCATIONS

V-1   MAXIMUM VELOCITY (TURBULENT)
V-2   AVERAGE VELOCITY (TURBULENT)
V-3   MINIMUM VELOCITY (LAMINAR)

Figure 1   Fluid velocity profile inside pipes and similar equipment.

components are in the fluid, they could collect disproportionately on the wall of the pipe or equipment, which would introduce errors in the cell count or other related analyses.

Another consideration for the proper design of the extended sample tube into the fluid is represented by points M and N. Point N has the end of the sample tube turned directly into the flow of the fluid. When slurry flows are encountered, solid particles could accumulate in the tube opening due to their momentum. When the sample point is opened, extra purging would be required to obtain a representative sample.

Insufficient purging could result in obtaining an inaccurate sample. Point M could have an opposite problem since the end of the tube is facing downstream. The momentum of any solids could pass by the tube opening and the sample could be deficient in solids, which would not be representative of the average fluid quality.

Another sample point design consideration is maintaining the sample withdrawal rate in the sample tube proportional to the average velocity in the bulk fluid. If the fluid is a slurry, then a slow withdrawal rate could cause separation of the solids and the sample would not be representative. Also, a slow withdrawal rate could alter the time relationship of the sample if rapid reaction kinetics occur.

The sample valves shown schematically in Figure 1 are flush-mounted ball valves or equivalent designs with a minimum dead zone to discourage bacteria growth and reduce flushing requirements. If dead zones are a significant factor in the specific process, a sample tube design such as Y or Z may be the best application. Maintaining turbulent flow in the sample tube system should offset the wall effect, although further evaluation would be needed using the spe-

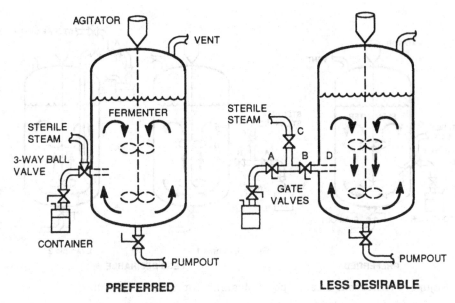

Figure 2 External sample tap connection designs for aseptic sampling.

cific biosystem during the equipment qualification and process validation stages.

Figure 2 shows two designs for piping and valving the aseptic sampling system from the bulk container to the sample container. The "preferred" model uses a flush-mounted three-way valve with minimum dead zone and a sterile steam connection for cleaning and sterilization between samples. The "less desirable" design has several features to avoid when aseptic samples are desired. First, gate valves are used which normally have relatively large dead zones that require additional cleaning and sterilization. Second, significant dead legs of piping are shown which will make purging and sterilization more difficult between points A to B, and B to D.

Figure 3 shows designs for the sample tubing inside the bulk fluid container. The "preferred" design shows a sample tube extension into the fluid where turbulent flow exists and the tube is angled about 15° downflow to prevent solids accumulation in the pipe between samples.

Example 1 uses the same principle for the tube extension at an angle sloping downstream. However, solids could accumulate in this pipe when the agitator is not operating. A better design may be to extend the tube horizontally into the fermenter. Example 2 shows

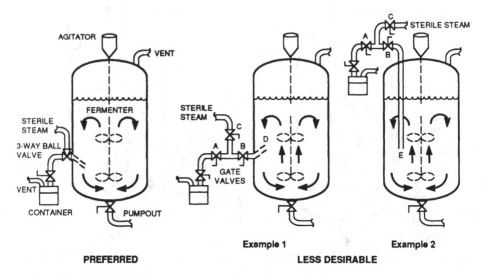

Figure 3   Internal sample point designs for aseptic sampling.

the sample pipe in the vertical position, which could have a longer
dead leg, and this may cause significant contamination entrainment
into the sample even if extended purging is used.

Containment of purges and venting from the sample container as
it is filled should be considered during the design of the specific
sampling system. If toxic or hazardous materials are expelled, they
should be piped to proper storage containers or locations for disposal. If necessary to protect the environment and operators, the
sampling system may need to be enclosed in a containment device
such as a glove box or hood system. An example of a protective
sample enclosure is shown later in Figure 4 [3].

Sampler and Container Design

Process conditions such as temperature, pressure, corrosivity,
particulate content, flows, viscosity, and other specific fluid characteristics are the prime consideration in the design of the sampling
system and sample container. Sample containers are usually available in glass, Pyrex, polyethylene, polypropylene, Teflon, and
stainless steel. Some containers are reusable and resterilizable,
while some are single use only. The sampling valves and piping
are normally stainless (316SS) with Teflon-seated valves where the
process conditions permit. If chlorides and higher temperatures
are encountered, more exotic materials of construction may be re-

quired, such as Hastalloy, Monel, and titanium which would increase the system cost. Adequate sample connections should be provided to remove the sample for analysis, especially if the sample contains solids or is viscous.

The process temperature may have to be reduced with cooling coils for safe handling and to stay within the design limits of the materials of construction. However, any cooling needs to be evaluated for potential solidification and high-viscosity effects. If the temperature is too cool, heat tracing or warming devices may have to be installed. Again, any process condition change would need to be evaluated for its effects on the sample quality and on the sampling system design.

The process pressure must be considered when specifying the sample valve and fittings. The sterilizing steam pressure should also be considered. The combination of pressure and temperature of the system design will establish the piping, fittings, and valve ratings to be used. Care is needed in sampling bulk fluids that are at higher pressures than will be encountered in the sample container. If the fluid is above its bubble point (temperature and pressure) in the sample container, significant vaporization may occur, which could be hazardous and could alter the composition of the sample, which may not be acceptable. When a fluid is under vacuum, a balancing vent may be required on the sample device, or a vacuum pump may be needed to draw the sample properly into the container within the prescribed time available.

When sampling slurries, the particle size of the solids must be considered when designing the tubing, valves, and other components to avoid plugging. Agglomeration of these particles and the effects of viscous fluids with the particles must also be considered. Special cleaning considerations should be included in the design for these types of slurries such as disassembly of the sampler to remove solids trapped in the valve mechanisms, cleanout ports to insert cleaning brushes or other devices to remove dried solids from the equipment walls, high-velocity spray jets to blast clean the walls of the sampling equipment, and back-flush connections to remove plugs of solid materials.

Sterilization design options may include in-place sterilization with the process equipment or removal of part or all of the sampler system for independent sterilization in steam autoclaves or ethylene oxide sterilizers. Figures 2 and 3 show one style of flushing sterile steam through a three-way valve to sterilize the sample path from the outlet of the process block valve into the discharge port for the sample container. Care must be taken to qualify this type of sampling system, since steam may not contact the surfaces of the valve near the process fluid entry, which could allow bacteria to propagate or die which could become pyrogenic (fever-causing) if the products are used for human injections. This entrained material could be difficult to remove by flushing and could contaminate succeeding samples. If this contamination is significant, some valves are commercially available which have steam jackets to heat the other dead zones for sterilization. However, pyrogen contamination may also remain, and the valve heatup may affect the bulk fluids in contact with the hot surfaces.

Purging considerations were previously discussed, and their significance depends on the nature of the biofluids and the design of the sampling system. Usually the goal is to purge the sample with minimum fluid to obtain the representative, noncontaminated sample. Several alternative purge design considerations are described here.

1. Use a three-way valve to divert the process fluid first to a purge line and then switch it to the sample container when the sample is deemed fresh and representative. It may be possible to divert the purged material back to a lower pressure point in the process so that the material is not lost.
2. Use a separate container to catch the purged fluid and then switch fluid to the sample container when ready.
3. Back-flush the process fluid inlet piping with sterile air or nitrogen or other inert fluids that will not affect the process.
4. Sampling devices can be designed to circulate the process fluid in an external loop by using either pumps or pressure differential and to trap a specified quantity of the fluid between two valves which can be discharged into the sample container.

Venting considerations for sample container filling are also needed, depending on similar criteria already described for the purging methods. If the purged vapors contain active microorganisms and are restricted from being vented to the atmosphere or local environment, the sample vent must be diverted to the proper disposal system. This diversion may include recycling the vapors to a lower-pressure system in the process which will not be affected by the returned vapors. Other alternatives for disposal include containers with neutralizing or deactivating chemicals, activated carbon drums, incinerators, slop storage drums for further processing, and deactivation of the biofluids and organisms.

Containment considerations have already been addressed. Standard containment designs are commercially available from fabricators of glove boxes and laminar flow hoods. However, custom designs, using established principles that can be validated, will probably be required. One example of an enclosed sample point is shown

Figure 4   Sampler device enclosure.   (Courtesy of DOPAK, Inc.)

in Figure 4 [3].   Another principle of containment design for consideration is to use a lower pressure in the containment chamber relative to the external environment to minimize leakage of biomaterials during sampling.   Also, catch basins can be used to control spills in the sample areas.   Offsetting this design philosophy is the need to slightly pressurize the enclosure to avoid outside contamination from entering the enclosure.   The overall merit of these two considerations must be compared for each specific application.

Several sampling system designs have been discussed.   Another commercial design involves the use of a sample bottle with a cap and septum that is inserted into a special sleeve below the sample valve until the septum is pierced by the needles at the top of the sleeve. When the sample is collected, the bottle is removed and the septum reseals the sample bottle.   Such a device is offered by DOPAK, Inc. (Plainsboro, NJ), and the principle is illustrated in Figures 5 and 6.

Another variation for multiple-sample collection is to install multisample devices that would be used for only one sample each during the batch.   This design would obviously require additional capital investment, although some savings per installation should be realized from replication of the sample device and from elimination of external sterilization equipment.   Sterilization of the samples would occur when the main tank system is sterilized before each new batch so

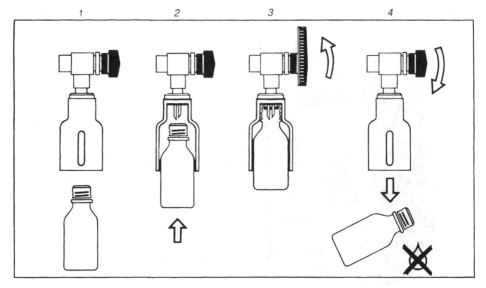

Figure 5   Aseptic sampler example.   (Courtesy of DOPAK, Inc.)

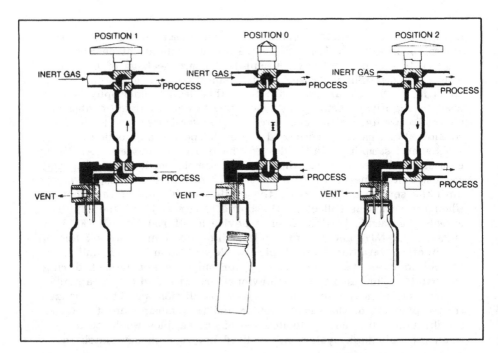

Figure 6   On-line sampler system.   (Courtesy of DOPAK, Inc.)

that the resterilization system could be eliminated. Care is needed to locate these multiple-sample taps so that equivalent sample quality can be obtained from each sampler and sample bias or errors will be avoided.

The trend toward automated multisample devices continues with aseptic sampling goals being within reach. These developments include multistation sampling and in situ probe sampling systems, which should be available in the near future. A novel sterile sampler has been produced by DuPont Biotech Systems (Wilmington, DE) for hollow-fiber culture ware in which a sterile tubing welder is used to splice a sample tube to a hollow fiber and to remove the sealed fiber tube of sample fluid.

B. Externally Removed Sample Design

One commonly used sample system is the direct sample tap with two block valves and a sterile steam connection between the two valves. This system was shown previously in Figures 2 and 3. The sample container can be inserted into a hood device at the end of the sample tube, as shown in Figures 5 and 6. Some systems used to fill the container use needle and septum connections; some use steam sparging as the sample is collected to avoid contamination entering the sample bottle.

Another method which has enjoyed traditional use is the rubber septum diaphragm and syringe system. Sometimes an antiseptic solution is absorbed on cotton and placed on the outside of the septum material, which sterilizes the needle as it penetrates the septum. As a sterile environment around this sampler system, a special glove box or other containment device can be provided. This containment chamber is the most practical method to contain aerosol release and spillage during sample collection. Special design conditions must be established for the septum diaphragm to withstand process conditions, especially pressure, when it is attached to the culture vessel wall.

Other custom designed sample valves are in use today. One such design from LH Fermentation Inc. (Hayward, CA) attaches a fixed-volume valve chamber to the culture vessel wall. The inlet to this chamber is sealed with a cam-operated plunger. After the sample is collected and the plunger seal is repositioned, the sample bottle or device can be removed and the sampler device resterilized.

Bioreactor and fermentation tank vendors also have customer design sample devices available for considerations. The Fluid Transfer Division of Lee Industries (Philipsburg, PA) offers a special aseptic ball valve design with steam tracing around the valve stem assembly, inlet flange, and end connections to enhance valve sterilization.

C. In Situ Sampler

The placement of an analytical device probe directly in the bulk
fluid without removing the sample for analysis probably provides
the most accurate sample. However, the analytical probe must be
capable of being sterilized and calibration-checked as needed during
the progress of the batch or continuous process. A primary concern is to position the probe in the proper location in the bulk fluid
where the fluid quality would be representative of the bulk material.
     As new analytical methods are developed, such as those dis-
cussed in this book, new ideas for sampler design will emerge to
overcome many of the current limitations of sampling devices.

IV. FABRICATION AND MAINTENANCE
GUIDELINES

The following guidelines are provided for the fabrication of sampling
devices and enhancement of procedures necessary to maintain aseptic conditions.

1. Use all welded construction where operations permit.
2. Use flanged connections instead of threaded fittings in aseptic
    material service.
3. Minimize dead zones in piping, fittings, and other equipment
    where the biosystem could develop contamination.
4. Select valves for biosystems that minimize dead space. Ball
    valves and custom-designed valves are commercially available for
    fabrication. Custom-designed rising-stem valves are available
    with small dead zones. These valves can be actuated manually
    or automatically as appropriate for the process and its control
    strategy. These actuators can be pneumatic, hydraulic, or
    electric. Consider installing flush bottom valves on tanks to
    minimize dead legs of material in the inlet to the valve.
5. Locate sample lines on or in vessels to avoid solids and other
    unwanted accumulations.
6. Avoid rough surfaces, cracks, crevices, and voids, where in-
    fections can occur. Grind welds smooth. Welding quality is
    most important to avoid pits and recesses for contaminants to
    accumulate or develop.
7. Select materials of construction for operating conditions. If
    process conditions are changed, an engineering review should
    approve the changes and define required revisions in the proc-
    ess equipment to meet the new conditions. The changes should
    be documented in the written procedures.
8. Equipment lubrication methods and lubricant selection should be
    tailored to the process. Silicone-based lubricants can replace

some petroleum-based lubes to avoid process contamination. Some types of rotating equipment have custom designs to avoid lubricant contact with aseptic fluids.

9. As with written operating procedures for the bioprocess, it is also important to use written maintenance procedures that are approved by engineering and operations management. All process changes should be reviewed with respect to impacts on the maintenance procedures and validation requirements. Proper changes should be incorporated into these documents.

V. SPECIFIC SAMPLING METHODS FOR BIOSENSORS

The general consideration for specific analytical methods were categorized in Section IIG as on-line, in-line, and off-line sampling requirements. Previous considerations for the design, operation, and maintenance of these sampling systems must be tailored to the level of aseptic sampling required for the specific biosystem.

Where current analytical methods use off-line samplers, it is often the goal that future improvements in the analytical devices may permit in situ probe installations to minimize the errors that improper sampling techniques can contribute to precision and accuracy of these analytical methods.

Trends in sampling and analytical testing during the past decade indicate that more in situ applications are needed and that the talents of the scientific community will be used to discover novel methods for improving these critical elements of process control for the Bioindustry.

For example, Sabelman [9] describes three approaches to automated sampling of sterile culture media by first transporting aliquots of sample fluids to a central bank of analytical instruments. A second method uses a series of analytical probes at the in situ sample point so that only electronic data are transmitted from the process fluids. The third method uses a proprietary sterility barrier and parallel-connected probes.

Since the sampling methods are process-specific, it is foreseen that multiple solutions to the sampling techniques will continue to emerge for the analytical methods discussed in this book, and custom designs will represent the majority of applications.

REFERENCES

1. U.S. Department of Health and Human Services, Food and Drug Administration, Guideline on Sterile Drug Products Produced by

Aseptic Processing, June, 1987 published by Center for Drugs and Biologics and Office of Regulatory Affairs, Food and Drug Administration, Rockville, MD.

2. Cooney, C. L. and Humphrey, A. E., Comprehensive Biotechnology, Vol. 2, Pergamon Press, New York (1985).

3. DOPAK Inc., DOPAK Liquid Process Samplers, DOPAK, Inc., Holland (1985).

4. Dwyer, J. L., Contamination Analysis and Control, Chaps. 3 and 6, Reinhold, (1966).

5. Scott, R. A. and Doemeny, L. J., "Design Considerations for Toxic Chemical and Explosives Facilities," Symposium, Division of Chemical Health and Safety. 194th Meeting, American Chemical Society, Washington, D.C. (1987).

6. Richards, J. W., Introduction to Industrial Sterilization, Chap. 8, Academic Press, New York (1968).

7. Block, S. S., Disinfection, Sterilization and Preservation, Chap. 41, Facilities for Control of Microbial Agents (1983).

8. East, D., Stinnett, T., and Thoma, R. W., "Reduction of Biological Risk in Fermentation Processes by Physical Containment," Paper presented at Meeting of the Society for Industrial Microbiology in Sarasota, Florida, Aug. 14–18 (1983).

9. Sabelman, E. E., "Strategies for Implementation of On-Line Analytical Sensors for Sterile Bioprocess Fluids," paper presented at ASME Winter Meeting 1988, Pro-Zooics Research, Menlo Park, CA.

3

On-Line Monitoring of Bioprocesses Using HPLC

BRUCE JON COMPTON *Bristol-Myers Squibb Co., Syracuse, New York*

I. INTRODUCTION

This chapter discusses recent progress made in adapting high-performance liquid chromatography (HPLC) analysis for on-line monitoring of industrial bioprocesses. It will become apparent that this field is in its developmental stage, and it is still not clear if HPLC will play a significant role in monitoring bioprocesses in a production environment or if it will be more useful in a process development role and, as such, have its greatest impact in pilot laboratories and pilot plants.

Traditional fermentation process development has used either batch or fed-batch operation and in-process monitoring to control, for instance, temperature, pH, agitation, aeration, pressure, and feed rates. Monitoring has involved dissolved oxygen, total nitrogen, pH, outgasses such as carbon dioxide and ammonia, and product measurements.

With the recent availability of inexpensive computing power and a general need to improve productivity, the next logical step in process development protocol is to study transient fermentation substituents (i.e., intermediates and degradation products) to gain insight into process-related biochemistry and rheology not easily studied in the laboratory. In a sense, we are attempting to bring the research laboratory into the pilot plant or to develop research-pilot-plant-sized bioreactors.

The main advantage of using HPLC over nonchromatographic methods, such as those based on spectroscopic and electrochemical measurements, is its ability to analyze multiple components in a complex matrix, such as fermentation broths or waste streams [1]. Its advantage over other chromatographic methods, such as gas chromatography, is its compatibility with nonvolatile solutions and thermally labile compounds such as proteins, antibiotics, etc. Finally, HPLC

has advantages over other liquid chromatographic methods, such as
thin-layer and paper chromatography, in that it can be automated
and, as will be shown, is a potentially rapid analysis tool.

HPLC has become well established as an analytical tool in the
last two decades because it actually encompasses a variety of operational modes [2]. The most widely used mode is reversed-phase
HPLC where, by definition, the mobile phase is polar relative to the
stationary phase. Typical systems are described later in this chapter. This mode is useful for separations based on the relative hy-
drophobicity of molecules. Many modifications to the reversed-phase
theme exist. For instance, the addition of ion-pairing agents to the
eluent (e.g., millimolar concentrations of t-butylammoniumphosphate,
trifluoroacetic acid, alkylsulfonic sulfonic acids) which combine by
ionic interactions with the analytes of interest, and cause selective
retention on hydrophobic stationary phases. A further variation on
the theme is the addition to the of coordinating metals mobile phase,
such as Cu(II), and chiral coordinating compounds (L-Proline) to
allow selective retention of compounds based on ligand exchange.

Other general modes of HPLC are size-exclusion chromatography
(SEC), which utilizes molecular sieving as a basis for selective retention; affinity chromatography, which uses biospecific interactions
for solute retention; normal-phase chromatography, which has a
stationary phase more polar than its mobile phase (e.g., silica gel
with a hexane mobile phase); and ion-exchange chromatography,
which uses an ionizable stationary phase for selective retention of
solutes with charge opposite to that of the stationary phase. It
should be obvious that HPLC is a remarkably versatile tool, since
its various modes are based on the physiochemical interactions which
represent the essence of modern chemistry.

In general, all classical liquid chromatographic modes have an
HPLC analog. For those unfamiliar with the distinction between
HPLC and classical chromatography, the fundamental difference is
that HPLC represents an attempt at optimizing the hydrodynamics
related to convectional processes and the kinetics related to molecular transfer processes. These attempts manifest themselves through
the use of well-characterized homogeneous column packings (usually
3- or 5-μm spherical diameter). Such packing has excellent mass
transfer characteristics and allows, because of its homogeneous nature, the construction of columns which have extremely uniform
beds. The small particle size of the packing and the requisite low
dead volume of the system means that high-pressure pumps (to
6000 psi) and low-dead-volume connecting tubing and detector flow
cells (less than 25 μL total) are required for its operation. The
operational conditions shown in Figures 1 and 2 are common for
state-of-the-art HPLC. The columns used for the separations were
all packed in house.

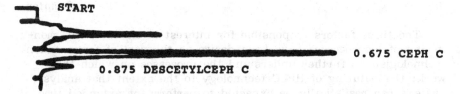

START

0.675 CEPH C

0.875 DESCETYLCEPH C

Figure 1 Fast HPLC separation of cephalosporin C and desacetyl-cephalosporin C using a 4.6 mm × 5 cm in-house packed 5-μm particle size APS-Hypersil column, 5 μL of a 20× diluted and filtered whole fermentation broth sample injected. Eluent 2200 parts CH$_3$CN, 120 parts MeOH, 80 parts acetic acid, 1600 parts water, flow rate 3 mL/min, 254 nm detection.

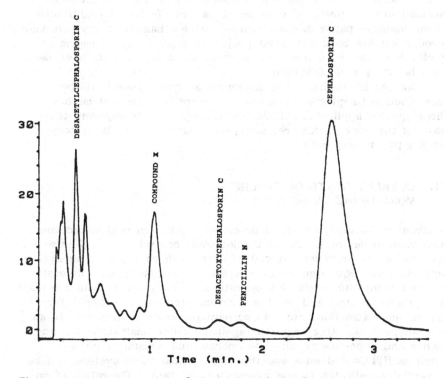

Figure 2 Fast separation of the major biochemical intermediates of cephalosporin C synthesis and the main decomposition product (designated compound X here). Conditions were 5 μL injection of a 20× aqueous diluted sample of whole broth onto a 4.6 mm × 5 cm Spherosorb ODS2 (5 μm) column, eluent 0.1% phosphoric acid with 5% methanol, 3 mL/min flow rate. Detection was by ultraviolet absorption 229 nm.

The three factors responsible for interest in on-line HPLC monitoring of bioprocesses have been desire on the part of fermentation technologists to further understand the processes on which they work; the maturing of HPLC technology to the extent that analysis systems can realistically be expected to perform for extended time periods (greater than 100 hr); and the maturing of the HPLC instrument market to such an extent that market diversification on the part of instrument manufacturers is sought.

Some of the early work in this area is described in a recent review on fermentation instrumentation and control [3] and on-line bioprocess monitoring using mass spectrometry [4]. Two laboratory-scale systems have been described for feedback control of substrate during ethanol production [5,6]. Other laboratory-scale work concerned with penicillin and cephalosporin production has been described in a comprehensive series of papers [7–9]. An additional comprehensive paper dealing with lab-scale ethanol and erythromycin production has been published [10]. This list is by no means complete, but illustrates that most systems described to date have been on laboratory scale fermentors.

Commercial on-line chromatographic analysis systems will be described in Chapter 4. The manufacturers of these systems have little specific applications literature to supply. It is apparent that most of the work in this area being done outside of the laboratory is of a proprietary nature.

II. CURRENT STATE OF ON-LINE
   MONITORING USING HPLC

Current HPLC analyses are done either off-line in centralized control laboratories or at-line at a pilot plant control center (Figure 3). Centralized laboratories evolved as arenas where a core of analytical expertise and equipment could operate to the benefit of a diversity of pilot-plant and production operations. Analytical results are used for process control and to allow various production sections (fermentation, extraction, etc.) to account for their contribution to an overall process. One argument against on-line analysis is the expense and expertise required to operate and maintain equipment such as HPLCs and mass spectrometers, since these systems will be essentially dedicated to one process or one tank. Operation of on-line capital analytical equipment also often involves multidepartmental effort, something which realistically is often more difficult to achieve than the technical aspects of projects.

The main advantage of on-line monitoring over more traditional off-line monitoring methods is that transient events with fast time

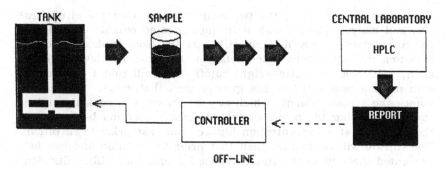

Figure 3 Schematic diagram of a generalized off-line analysis operation illustrating the use of a centralized laboratory facility.

constants can be monitored. For instance, monitoring bioreactor responses to operation perturbations such as changes in agitation and tank pressure can be done dynamically. If experiments are conducted during a linear production phase of a process, instantaneous production rate constants can be determined for a variety of conditions, potentially allowing optimization of process variables for a production phase separate from early bioreactor growth phase.

HPLC systems are, in general, sensitive enough (often submicrograms on-column) and flexible enough to allow direct clarified sample introduction. Sample clarification involves either centrifugation or filtration through a 0.45-μm filter. Sample dilution is necessary when sample viscosity or concentration exceeds filter or column capacity, respectively. Figure 4 illustrates a generalized on-line HPLC analysis system.

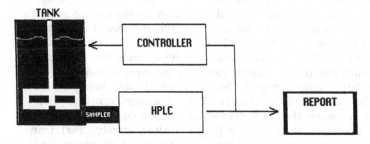

Figure 4 Schematic diagram of a generalized on-line analysis operation for direct process control.

For on-line analysis, the two sampler types used are ultrafiltration and dialysis [3—10] with units incorporated outside or inside of the bioreactors. Ultrafiltration (UF), as the name implies, involves filtration through filters which retain at the molecular level. As such, a 100,000-molecular-weight cutoff filter will retain compounds with nominal molecular weights greater than that value. In this instance the stream diluent, which may be water, will pass through the filter leaving the high-molecular-weight compounds behind in the retentate at a concentration higher than that prior to filtration. The filtrate will be cleaner than that prior to filtration and can be presented directly to the HPLC system for analysis. Ultrafiltration usually occurs under pressure to force mass across the membrane, and is subject to membrane fouling as the membrane becomes coated with a particulate or gel film, depending on the sample. The optimum operation of a UF system usually involve a configuration which causes high shear across the membrane by application of tangential flow [7—10].

Dialysis also involves filtration at the molecular level, but in this case the driving force for filtration is osmotic pressure since the dialysis configuration involves a concentrated sample stream on one side of a membrane and a clean water or buffer stream on the other. Solute which passes through the membrane partitions is diluted. Dialysis is less subject to membrane fouling, but can lead to erroneous results if operated in a nonequilibrated fashion.

Besides sampling, long-term unattended operation of an HPLC system and a means of using the systems results are both necessary. Dedicated chromatographic integrators with basic programming, hard copy, and hex ACS II transmission via RS-232C are obtainable from a variety of instrument manufacturers for under $4000. The main problem with implementing HPLC control of bioreactors is the general cost and complexity of an HPLC system designed for an industrial environment, since most systems to date are for laboratory settings. A system from Waters ProMonix (Milford, MA), with full support costs over $100,000. This organization has attempted to construct a reliable instrument by incorporation of a high-pressure syringe pump rather than more expensive dual-piston reciprocating pumps and eluent recirculation. These steps should increase reliability but at the expense of system flexibility in pressure limit and analysis time, since syringe pumps have long cycle times for refilling while reciprocating piston pumps run in a continuous mode.

On the latter point, recent trends in HPLC have been toward fast analysis (seconds to minutes). This is a significant development, since it means that, by averaging many analyses, highly accurate and precise results can be found. Alternatively, fast transient bioreactor responses, such as occur following substrate additions, can be followed [10]. Figure 1 shows a rapid separation of cephalosporin C from desacetylcephalosporin C and other broth constituents. Analysis was completed in 1 min. Figure 2 shows separations of cephalosporin in whole broths under conditions which detect multiple components of interest. The cephalosporin C separation is interesting, since it allows determinations of the complete kinetics of this antibiotic. Penicillin N, desacetoxycephalosporin C, and desacetylcephalosporin C are the three biochemical precursors to cephalosporin C, while compound X is the main decomposition product of the antibiotic.

III. FUTURE PROSPECTS

The maturation of the HPLC market and technology and the commercial availability of systems mean that outside support for on-line HPLC analysis, while in its infancy, is available. Consequently, systems development and support can be delegated to outside sources at a cost. As with new commercial ventures, a period of close synergistic interaction between suppliers and users is required before commercial systems will be fully functional and turnkey.

The maturation of HPLC technology is certainly the technical aspect of the subject, which is of premier importance. For instance, the use of valve switching methods will allow automatic sample concentration, dilution, and column and filter back-flushing. Fast HPLC will extend the use of the method to monitoring fast transient signals or, as a trade-off, signals which are averaged to give greater accuracy.

The area of greatest concern, as is always the case when doing analyses, is obtaining representative samples. Overcoming the seemingly mutually exclusive problems of increasing filter surface while decreasing sampler holdup volume needs attention from filter suppliers. Even systems based on filter back-flushing suffer from clogging and contamination [1]. The good news is that on-line samplers for HPLC will be similar in requirements to those designed for on-line spectroscopic measurements, mass spectrometry, and automated wet-chemical techniques. Consequently, while each bioprocess is unique in its rheology and biochemistry, one sampler design for each process should be enough for many analysis techniques.

REFERENCES

1. Giddings, J. C., <u>Dynamics of Chromatography: Volume I. Principles and Theory</u>, Marcel Dekker, New York (1965).

2. Snyder, L. R. and Kirkland, J. J., Introduction to Modern
    Liquid Chromatography, Wiley, New York (1974).
3. Carleysmith, S. W. and Fox, R. I., in Advances in Biotech-
    nological Processes, Vol. 3 (A. Mizrahl and A. L. van Wezel,
    eds.), Liss, New York (1984), pp. 14–16.
4. Heinzle, E., in Advances in Biochemical Engineering/Biotech-
    nology, Vol. 35 (A. Fiechter, ed.), Springer-Verlag, Berlin
    (1987), pp. 2–41.
5. Mathers, J. J., Dinwoodie, R. C., Talarovich, M., and Meh-
    nert, D. W., Biotechnol. Letters, 8, 311 (1986).
6. Dinwoodie, R. and Mehnert, D. W., Biotechnol. Bioeng., 1060
    (1985).
7. Moller, J., Hiddessen, R., Niehoff, J., and Schugerl, K.,
    Anal. Chim. Acta., 195 (1986).
8. Niehoff, J., Moller, J., Hiddessen, R., and Schugerl, K.,
    Anal. Chim. Acta., 205 (1986).
9. Bayer, T. H., Herold, T. H., Hiddessen, R., and Schugerl,
    K., Anal. Chim. Acta., 190:213 (1986).
10. Dincher, A. K., Manohar, K., Skea, W., Ryan, M., and
    Kierstead, T., in Developments in Industrial Microbiology,
    Vol. 25 (C. H. Nash and L. A. Underkofler, eds.), Society
    of Industrial Microbiology, Arlington, VA, pp. 603–611.

4

High-Performance Liquid Chromatography

KENNETH J. CLEVETT   *Clevett Associates Inc., Watchung, New Jersey*

I. HISTORY OF CHROMATOGRAPHY

Chromatography is a physicochemical process whereby a sample containing several chemical compounds may be separated into individual components of the mixture. It works in a discontinuous manner. A small sample is taken and the individual components of the mixture are retained on the chromatographic column to different extents, as if they were distilled off one by one. Because of its nature, the separation process usually takes about 5 to 20 min (known as the cycle time). As the individual components emerge (elute) from the column, each is measured by a detector, the output of which is calibrated to be proportional to the concentration of that component in the original sample.

The invention of chromatography is generally attributed to the Russian biochemist Tswett, who was interested in the green coloring matter found in plants (chlorophyll). In 1903 he reported the separation of different plant pigments, which were visible as colored bands when a solution of chlorophyll was washed by a suitable solvent through a tube containing an adsorbent such as powdered chalk. In a paper published in 1906 [1], Tswett named this technique "chromatography" (literally, color writing), although he appreciated that the technique could be used for separating colorless substances as well. Further development of the technique took place between 1903 and 1910. The technique discovered by Tswett was rarely used until 1931, when the German scientists Kuhn, Lederer, and Winterstein published papers [2,3] on the chromatography of carotenoids. The value of the method was fully realized in these papers, the method thereafter was generally accepted, and the number of applications in a variety of different fields increased rapidly.

Four years later, Adams and Holmes [4] published work on ion-exchange chromatography, which led to the preparation of the first synthetic organic ion-exchange resins. This was followed in 1941 by the introduction of partition chromatography by Martin and Synge [5], resulting from their work on the separation of acetylated amino acids from hydrolysates by extraction. The method was further developed by Martin and his coworkers to a special form of the technique known as paper chromatography. For their highly successful contribution to the field of biological and medical research, Martin and Synge received the Nobel Prize in 1952.

The development of chromatographic techniques, however, did not stop here. The fundamentals of gas adsorption chromatography had been known since 1936 [6], and in 1952 James and Martin [7,8] introduced the technique of gas chromatography and presented the results of their work at the Analytical Chemistry Congress at Oxford, England in that year. One of the characteristic features of the method described was the very small samples that were used for the estimations. The simplicity and analytical power of the technique were immediately recognized. Because of its promise, gas chromatography received much attention and its development was very rapid. Since 1952, the growth in both the theoretical and practical aspects of the technique has been tremendous. Not only has it provided a simple solution to many complex laboratory analyses, but it has provided an efficient method for on-line monitoring and control of industrial processes.

II. EVOLUTION OF LIQUID CHROMATOGRAPHY

A. Background

Although liquid chromatography (LC) has been known even longer than gas chromatography (GC), until quite recently its use has been largely confined to the laboratory, where it is a very well established and extensively used analytical technique. Process liquid chromatography (PLC) is probably now at about the same stage of development as process gas chromatography (PGC) was in the early 1960s.

In the last 20 to 25 years, there has been a phenomenal development in the technique known as high-performance liquid chromatography (HPLC). Although it is very difficult to compare the evolution of the various analytical techniques used in the laboratory, the growth of HPLC is considered by many scientists to be unparalleled in the field of analytical chemistry, even compared with that of gas chromatography.

The term HPLC, as applied to the present-day technique, should not be construed to imply that earlier LC did not exhibit high performance. On the contrary, all of the pioneering scientists in this field, in Germany, Hungary, England, and the United States, carried out separations hitherto impossible by using other analytical techniques. Therefore, even by present-day standards, the earlier separations can certainly be classed as "high-performance" LC. In fact, the HP in the name originally stood for "high pressure," since the change from atmospheric conditions and gravity flow to pumping systems and high pressure seemed to be the most significant difference between the old and new techniques. However, it was soon recognized that high pressure was not the main characteristic of the new technique. It was proposed by Csaba Horvath in 1970 that the name be changed to high-performance liquid chromatography, and this was almost immediately accepted on an international basis to describe modern liquid chromatography.

B. Advantages of Present-Day LC

There are four basic areas where present-day LC performance is superior to that obtained by the pioneering scientists in this technology.

1. Speed—Separations that took 1 or 2 hr to perform 40 years ago can now be done in minutes.
2. System performance—Present-day LC is simpler, more precise, and reproducible, mainly due to improvements in column technology, detector design and sensitivity, and stability of mobile phase flow control.
3. Sample size—Whereas classical LC was a semipreparative laboratory method, HPLC is a microtechnique, requiring only very small samples.
4. Theoretical basis—While classical LC was essentially empirical in nature, the development of HPLC was based on theoretical principles that led to improvements in the technique.

III. TYPES OF LIQUID CHROMATOGRAPHY

The chromatographic process is based on the fact that the individual components of a sample mixture, under identical conditions, are distributed to different extents between two phases, stationary and mobile. The stationary phase is a fixed bed of solid particles which may or may not be covered with a liquid coating. The mobile phase, or carrier, is a fluid medium that transports the sample mixture past the fixed bed or stationary phase. As the mobile phase permeates through the stationary phase, components that are more soluble or have greater affinity for the stationary phase move through the column at a slower rate than those components that favor solubility in the mobile phase. The net effect is that the various components are separated into individual elution bands for analysis.

The various types of chromatography are classified by the nature of the <u>carrier</u> (or mobile phase) and the nature of the <u>static medium</u> (or stationary phase) as follows:

Mobile phase	Stationary phase	Type
Gas	Liquid	Gas-liquid chromatography (GLC)
Gas	Solid	Gas-solid chromatography (GSC)
Liquid	Liquid	Liquid-liquid chromatography (LLC)
Liquid	Solid	Liquid-solid chromatography (LSC)

Even within a type of chromatography, as outlined here, different separation techniques can be applied. Some of these techniques have become so widely used that additional subcategories have evolved. In the case of liquid chromatography, these subcategories are described next.

A. Liquid-Liquid Chromatography

Liquid-liquid chromatography is partition chromatography in which component separations are achieved by partitioning of the sample between a liquid carrier and a liquid stationary phase that is coated onto a solid packing material. For stability, the liquid stationary phases that are now used in LLC are chemically bonded to the packing material, thereby eliminating the potential problem of the liquid carrier washing the stationary phase off the solid packing, and providing a column that meets the long-term stability requirements for on-line analysis.

Liquid-liquid chromatography is further subdivided into normal-phase and reversed-phase systems. Normal-phase systems use a carrier that is less polar than the stationary phase, whereas reversed-phase systems use a carrier which is more polar that the stationary phase. Reversed-phase partitioning systems have become so popular that this technique is often referred to as "reverse-phase LC"; roughly 75% of all laboratory separations involve reversed-phase partitioning. One of the main reasons for its popularity is that many organic compounds can be separated with water as the major carrier solvent. However, the analysis time is often long-er than for other forms of partitioning chromatography.

The reversed-phase system is used extensively in the separation of organic compounds, particularly hydrocarbons that differ only in their carbon number. In general, paraffinic hydrocarbons require an organic or nonaqueous carrier for separation, whereas aromatic hydrocarbons can be separated using aqueous blended carriers with higher water content.

When the separation involves highly polar compounds, reversed-phase systems may be used with an additional "modifier" in the carrier. This causes either ion pairing or ion suppression of the polar sample. In the case of ion pairing, the modifier is also ionic in nature, and combines with the sample to produce a less ionic paired species. Once ion pairing has taken place, the retention time of a particular component is more dependent upon the hydrocarbon portion of the molecule.

In ion-suppression systems the modifier added to the carrier controls the pH and causes ionic components of interest to exist mostly in their neutral form. Once this occurs, standard reversed-phase techniques can be used.

B. Liquid-Solid Chromatography

Liquid-solid chromatography, or adsorption chromatography, is the oldest form of chromatography (e.g., Tswett's work on plant pigments). However, it was not until the introduction of modern high-performance packing materials in the late 1960s that this form of chromatography was able to provide efficient and timely separations and became popular. Adsorption chromatography involves the competition of the sample components between the active adsorption sites of the packing material and the solvent molecules. The most popular adsorption material is silica.

In general, adsorption chromatography is used for separation of organic compounds of intermediate molecular weight. Lower-molecular-weight organic compounds are usually best separated by gas chromatography. Adsorption chromatography is not suitable for high-molecular-weight organic compounds, nor is it suitable for ionic compounds which tend to "tail" appreciably. The most useful application of this technique is in the separation of organic compounds with different functional groups (e.g., aromatic isomers that differ only in the location of the functional group).

C. Ion-Exchange Chromatography

Ion-exchange chromatography (IEC), as its name implies, is used almost exclusively for the separation of ionic compounds, commonly in

an aqueous medium. The packing material in this case is typically
a highly permeable ionic resin, manufactured from the polymerization
of styrene and divinyl benzene with a suitable functional group.
However, the porous resins have gradually been replaced by the
new high-performance, bonded-phase ion-exchange resins, which
generally provide a more efficient separation and a more stable column for on-line use.

The greatest potential for on-line use of IEC lies in water quality monitoring for steam boilers, power plants, nuclear reactors,
and the like. The ionic species present in water are currently
measured on-line by a wide variety of different analytical instruments, usually one analyzer for each ion type. With IEC perhaps
all of the ionic species of interest can be measured with one instrument.

D. Size-Exclusion Chromatography

Size-exclusion chromatography (SEC) is also known by several other
names, including steric-exclusion, liquid-exclusion, gel filtration,
and gel permeation chromatography. However all of these terms
refer to essentially the same technique, namely, that sample components are separated according to their molecular size. Unlike the
other techniques of liquid chromatography, the separation process
in SEC is based on entropic rather than enthalpic interactions with
the packing material of the column. The retention mechanism is
therefore mainly dependent upon the molecular dimensions of the
sample and the physical characteristics of the packing. Molecules
that are smaller than the average pore size of the column packing
material will spend a significant amount of time within the pores of
the packing material, whereas larger molecules will spend less time
within the pores. This results in a molecular size or weight distribution chromatogram, where the larger molecules elute first and
the smaller molecules have longer elution times.

Size exclusion chromatography is an ideal technique for the
analysis (characterization) of many polymer streams, which is an important requirement in the petrochemical industry that has been
difficult to meet with alternative analytical technology. It is therefore the single most important application for on-line liquid chromatography at the present time.

Industrial polymers are generally produced from chemical reactions between relatively simple organic compounds, in which each
reactive compound, or monomer, behaves as an individual unit for
a "building-block" process that results in the formation of a specific
polymer. Most specific polymers are mixtures of a range of different molecular species, where many of the characteristic physical

properties of the polymer are related directly to the degree of distribution within the mixture. SEC provides a method for characterizing the polymerization process by means of a molecular-weight distribution (MWD) analysis.

The following equation defines the most commonly used parameters for characterizing a given polymer:

$$M_b = \frac{\sum N_i M_i^b}{\sum N_i M_t^{(b-1)}}$$

where N_i is the number of species with a molecular weight of M_i.

When $b = 1$, the equation yields the number-average molecular weight (M_n), which can be related to colligative properties of the polymer, such as vapor pressure and osmotic pressure. When $b = 2$, the equation yields the weight-average molecular weight (M_w), which can be related to the light scattering behavior of the polymer in solution and to properties such as mechanical strength. In addition, the ratio of the two averages (M_w/M_n), known as the polydispersity, is of interest, since it describes the breadth of the molecular distribution. A polydispersity of 1 indicates a polymer with a narrow distribution of molecular weight.

IV. ON-LINE PROCESS HPLC

A. Introduction

Although several process chromatograph manufacturers have shown an interest in PLC and a number of prototypes have been produced, to the author's knowledge there are only three companies currently manufacturing on-line liquid chromatographs for industrial process applications, namely, Applied Automation Inc., Dionex Corporation, and the Waters Chromatography Division of Millipore Corporation.

B. Commercially Available Equipment

Applied Automation Inc.

Applied Automation (AA) installed its first prototype process LC in 1974, and the first publications covering on-line experience appeared in 1976 and 1977. AA now reports over 50 units on order or in operation in the United States. Many of these installations are in applications involving the analysis of the molecular-weight distribution of polymers, utilizing the SEC technique.

Although the requirements for on-line HPLC were first incorporated successfully into the Optichrom 102 chromatograph system,
the introduction of the microprocessor-based Optichrom 2100 system
in the early 1970s provided integral facilities for the mathematical
computations required for polymer characterization. Quite recently,
Applied Automation's on-line LC has been incorporated into the
Optichrom ADVANCE chromatograph system. This is a truly distributed analyzer system based on the data hiway concept, the capabilities of which are shown in Figure 1.

The Optichrom LC, shown in Figure 2, is similar to the Optichrom GC in layout. At the top is the electronics section, which
incorporates a single-board microprocessor (which controls the LC
operation) and the fluids control panel. The center section, the
oven, houses the chromatographic components such as column(s),
switching valves, and detector. The sample-handling components
are mounted in the bottom section. The portable service panel
(PSP) plugs into and magnetically attaches to the electronics unit
and allows communication with and data display from any device on
the ADVANCE hiway. The ADVANCE system has the option of a
centrally located IBM PC/AT for postanalysis data processing,
graphics, and data storage.

The Optichrom LC electronics has all the capabilities of the GC,
including extensive internal self-diagnostics, interfaces for host
computer, printer, etc., and multistream, multicomponent analysis.
Three detectors are available: refractive index, UV—near-IR optical absorbance, and dielectric constant types. The solvent carrier reservoir is designed for up to eight weeks of continuous operation and is usually located adjacent to the analyzer.

Dionex Corporation

Dionex has been involved in industrial ion analysis for over a
decade in areas such as power generation, chemical processing,
fermentation, plating, pulp and paper and semiconductor manufacture, and probably has the most experience in this field. The
Series 8000 on-line ion analyzer was introduced about three years
ago. The latest, second-generation instrument, the Series 8100,
was first introduced in 1987 and incorporates a number of new
features and improvements based on experience with the Series
8000 in industrial process monitoring.

The Series 8100 ion analyzer, shown in Figure 3, is mounted in
a NEMA 12 enclosure that can be equipped with an air conditioner
if necessary for operation in a hot environment. The chromatograph, which can be utilized for ion chromatography, liquid chromatography, and/or flow injection analysis, is completely operated
and controlled by a Hewlett-Packard 310 computer system, shown

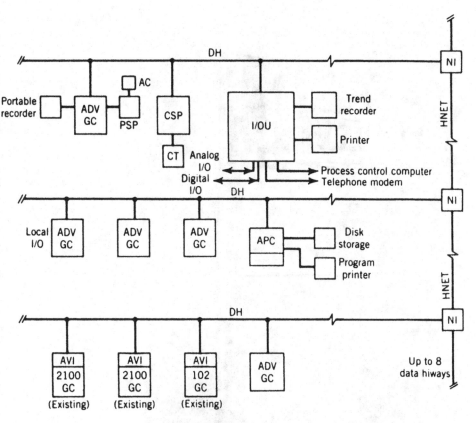

Figure 1   Optichrom ADVANCE System. (Courtesy of Applied Automation Inc.)

in Figure 4. An external sample stream selector is provided for
selection of 6, 12, or 18 process streams, housed in a NEMA 4 enclosure. The analyzer is modular in design for ease of maintenance
and for modification if different analyses are desired. Analysis
reagents are stored at the base of the enclosure for ease of re-
plenishment.

The computer system may be located near the analyzer (within
12 ft) or remotely at distances up to 1250 m when HP 3724A extend-
ers are used. The system provides data trending graphs, con-
densed reports on a shift, daily, or weekly basis, and high/low
alarm limits for each measurement. Capability is also provided for
output signals for the control of pumps, switches, remote alarms,

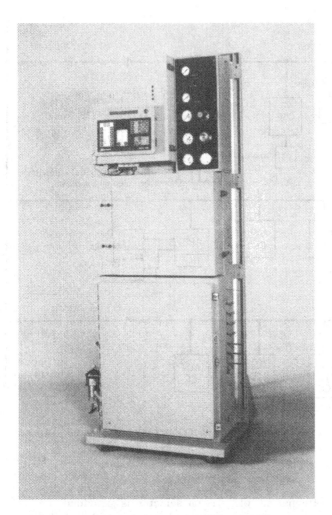

Figure 2 Optichrom ADVANCE Liquid Chromatograph (Courtesy of Applied Automation Inc.)

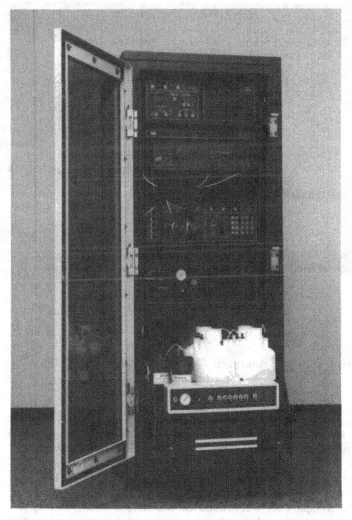

Figure 3 Series 8100 Ion Analyzer (Courtesy of Dionex Corporation).

Figure 4   Series 8100 Ion Analyzer Controller (Courtesy of Dionex
Corporation).

etc.   The analyzer system can also be interfaced with other computers using an RS-232C serial data link.   In addition, the analyzer
can be configured to take measurements other than ion determination, such as pH.

Some of the improvements incorporated in the Series 8100 include the following.   The Dionex Model GPM eluant pumps replace
the Model APM pumps that were used in the Series 8000, and allow
the user to automatically dilute eluant concentrates with pure water
once or twice per month.   (The Series 8000 required makeup of
fresh eluants twice a week.)   The system allows the suppressor re-
generant to operate without attention for a period of weeks.   Drip
trays are included between modules, which catch any small leaks
and keep them from damaging the electronics.

The chromatograph is capable of measuring the following ions in
concentrations ranging from a few parts per billion (ppb) to higher
percentages:

Cations—sodium, calcium, magnesium, ammonium, hydrazine
Anions—chloride, sulfate, nitrate, phosphate, hypophosphite,
    phosphite, silica
Heavy and transition metals—iron, copper, lead, nickel, cobalt and
    others
Organic compounds—organic acids and a variety of other organic
    ions

The most sensitive measurement ranges for some typical ionic species in solution are as follows:

Chloride	0.1—50 ppb
Sulfate	0.1—50 ppb
Sodium	0.1—50 ppb
Ammonium	1.0—100 ppb
Silica	1.0—100 ppb
Iron, copper, zinc, nickel	1.0—100 ppb (approx.)

Millipore Corporation

The Waters Chromatography Division of Millipore Corporation
recently introduced the Promonix range of on-line HPLC analyzers,
which are designed to perform on-line analyses based on ion, gel
permeation, or other LC techniques. The analyzer has the capability for analysis of up to 15 process streams, with automatic sample
conditioning prior to each analysis. Typical cycle times are quoted
as 5—12 min. The analyzer consists of the chromatographic compartment, power and air purge compartment, and electronics compartment, as shown in Figure 5. The chromatographic compartment
contains all analysis and fluid-handling components, such as pumps,
valves, and diluter/concentrator, in an environmentally controlled
chamber. The column(s) and detector are mounted in individual
temperature-controlled compartments. UV/Vis, refractive index, and
conductivity detectors are available. The solvent reservoir may be
located inside the lower compartment or external to the analyzer.
The power and purge compartment contains the power switch and
air purge module, which purges and maintains a positive pressure
in the analyzer for operation in National Electrical Code (NEC)
Class 1, Group D, Division 2 hazardous locations.

The electronic compartment contains all microprocessor-based
components for control, data acquisition, and storage. Plug-in
modules are used for ease of servicing. A keypad allows analyzer
operating parameters to be set, and selection of status and system

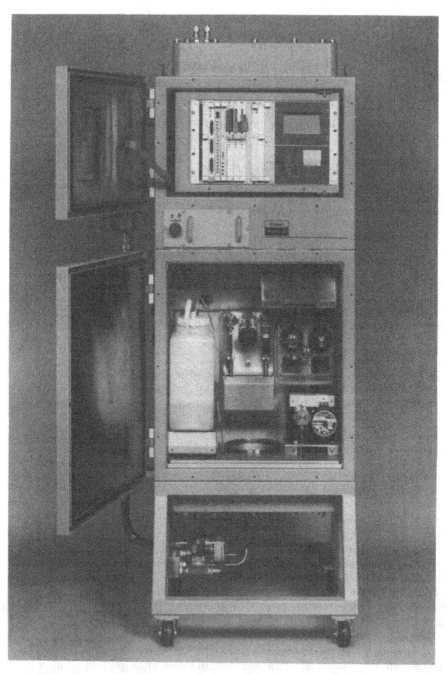

Figure 5 Promonix On-Line HPLC Analyzer (Courtesy of Waters
Chromatography Division, Millipore Corporation).

diagnostics, operational parameters, trends, results, and chromatograms. The Promonix HPLC analyzers are designed as standalone devices, or they can be data linked to personal or mainframe computers.

C. Considerations for On-Line Application of HPLC

As with all process analyzer applications, several points must be considered when this technology is applied on-line in the process industries. The following are considered to be pertinent in the application of HPLC.

Liquid Chromatography Versus Gas Chromatography

The on-line PGC was the first instrument to provide the capability of measuring process stream composition by separation and measurement of individual components. PLC expands this capability to include component separations previously considered impossible by GC or other analytical techniques. Therefore PLC should be looked upon as complementary to rather than competing with PGC, because it provides additional capability in areas where PGC cannot provide an analysis.

Note that there is considerable overlap between these two chromatographic techniques. However, in view of the more advanced development of PGC at the present time, there is a general preference for PGC over PLC on the part of potential users. This apparent conservatism is probably the reason why PLC has not achieved its full potential to date. In addition, for field operation and maintenance of PLC, some form of specialized training is required, whereas most analyzer engineers and technicians are now familiar with PGC.

Laboratory LC Versus PLC

As mentioned earlier, liquid chromatography is now an internationally recognized and powerful analytical technique for use in the laboratory. PLC, however, is still in an early stage of development. As with the development of almost all on-line process analyzers, the reasons for this are best illustrated by looking at the differences between laboratory and PLC.

First, the laboratory instrument is designed for manual operation in a safe, stable environment, whereas the process instrument must be designed to operate automatically and unattended in a relatively harsh environment, with wide variations in ambient temperature. It must be weather-resistant and meet the requirements for installation in locations often classified as hazardous for electrical equipment.

Second, the laboratory instrument is usually designed for versatility, that is, frequent changes in operating parameters and programming, and the capability to perform analyses on several different samples. Conversely, the process instrument is usually designed for a dedicated sample analysis and for stable and reliable operation in a field environment.

Third, the emphasis in process LC is to provide the fastest possible analysis utilizing multicolumn, multivalve technology. In addition, more emphasis is placed on solvent selectivity and optimization than on laboratory LC.

Fourth, PLC requires a relatively complex sample-handling system to transport and condition the process sample so that it is compatible with the analyzer requirements. In contrast, the laboratory instrument requires less complicated sampling facilities, although this is almost certainly not true for potential biotechnology applications.

Another difference is in the life expectancy of the columns used. The life-span of laboratory columns is reasonably estimated at about 1000 injections. In contrast, a PLC might be expected to complete 1000 injections in under a week of operation. Therefore at least a 10-fold increase in column life expectancy is desirable for PLC. The reliability, service factor and performance of an on-line analyzer should be as high as possible, so that equipment maintenance can be minimized. It is therefore important that the columns used in PLC be capable of performing to these expectations. PGC manufacturers usually guarantee their columns for a year's operation. In the case of PLC, and especially in the bioprocess area, we are probably treading new ground here, but the same rules will apply.

One very significant factor that must be considered is the lower level of expertise in LC technology of process analyzer maintenance personnel, compared with personnel responsible for operation and maintenance of laboratory LC instruments. This is to be expected, since PLC is a young technique and very little experience is available at present. Therefore, it is essential that training of field maintenance forces be given top priority.

Finally, the current cost of PLC is much higher than laboratory LC, since in most cases the cost of PLC includes both hardware and applications engineering, whereas the laboratory LC is usually sold as hardware only and the applications engineering is done in-house by the user. A well-designed sample-handling system required for the PLC is usually a very significant cost item and may be as complex and costly as the analyzer itself.

The Importance of Sample Handling

As with all on-line process analyzer applications, the importance of correct design of the sample handling system cannot be overemphasized. Industrial experience over the past two decades has shown that the sample-handling system is the source of at least 80% of the problems with on-line analyzer installations. The process analyzer usually will perform acceptably and reliably as long as it is provided with a timely, reliable, and representative sample of the process stream that is compatible with its operation.

Poor design of the sample handling system can lead to major problems in the analyzer itself, resulting in unreliable operation and loss of confidence on the part of operations personnel in the total analyzer system. For example, poor filtration in the sample conditioning system will result in plugging of the analyzer and associated components with particulate matter, requiring excessive down time for maintenance. Any time that the analyzer is out of service for long periods, there is a possibility that confidence will be lost.

The net result is lack of confidence in process analyzers generally. This problem was generated in the early 1960s by lack of experience and understanding, and is still being faced today. As is true in many other fields, reputations are built up over long periods but are lost very quickly!

It is safe to say that the majority of the conventional industrial applications of process analyzers have now been satisfied, and we are beginning to tackle those applications which involve difficult sampling and complex analysis. Therefore, the role of the sample handling system is no doubt going to become more important. For instance, the example of MWD analysis, mentioned earlier, which is an important application of size-exclusion liquid chromatography, involves complex sample handling and complex mathematical manipulation of the data generated. A typical sample handling system for on-line GPC by Applied Automation Inc. is shown in Figure 6. In the case of the potential applications of HPLC in biotechnology, there is every reason to believe that we are going to be faced with major challenges in the sample handling area.

V. DATA ACQUISITION FOR PROCESS CONTROL AND OPTIMIZATION

The process analyzer has now become an important component of advanced control systems, and its role as compositional sensor has added a significant dimension to overall process control strategy. Accessibility of reliable compositional data has resulted in improved regulatory control, and has also made a major contribution to proc-

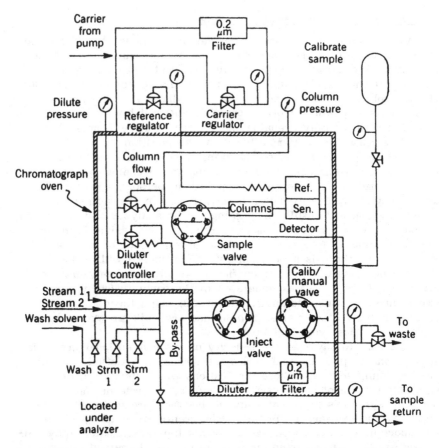

Figure 6 Typical Sample Handling System for Molecular Weight Distribution Measurement Using SEC. (Courtesy of Applied Automation, Inc.)

ess optimization in terms of improved quality and other advanced control applications. There is, therefore, a very real justification for providing, reliable, up-to-date analytical information for use in process control.

The successes attributable to the use of analyzers in industrial process control applications include savings in production, product giveaway, operating manpower, and energy conservation. These savings can be considerable. For example, there are individual analyzer installations in the petroleum and petrochemical industries that have generated savings of over $500,000 per year.

Traditionally, process analyzer data are transmitted to the process control system as analog (4—20 mA DC) signals in the same manner as conventional flow, pressure, and level signals. Some manufacturers of pH and conductivity analyzers went a step further with the introduction of the "two-wire" transmitters that look, from the instrumentation system end, exactly like a conventional field flow or pressure transmitter. The current loop may be powered by a 24-VDC supply, thus eliminating the need for a separate power supply at the transmitter.

A. Microprocessor-Based Analyzer Systems

The advent of the microprocessor in the mid-1970s had a profound effect on the development of instrumentation systems and on the philosophy of process control. This remarkable device has also had a major impact on the design of process analyzer systems. In fact, the pace of development in this area has been so rapid that it is difficult to keep up to date on new microprocessor-based process analyzers entering the marketplace.

One of the major focal areas has been the development of microprocessor-based GC systems, covering single GC controllers and systems capable of supporting multiple GCs. The Applied Automation ADVANCE system is an example. The advantages of such systems over the traditional analog systems may be summarized as follows.

- Multiple GCs can be controlled from one programmer.
- GC programming functions can be carried out more precisely.
- The microprocessor has the capability to perform extensive internal self-diagnostic routines to monitor system performance.
- The programmer can be interfaced with a printer to provide analysis reports, alarm reports, system status reports, and so on.
- The multianalyzer system can be connected by a single digital data link to a host computer, thus simplifying the communications requirements.
- Analog signals can be provided, if necessary, for direct transmission to the process instrumentation system.
- The power of the microprocessor can be used to perform special calculation routines, such as BTU, specific gravity, time-weighted averages, and so on. It can also carry out more sophisticated peak resolution routines such as perpendicular drop, tangent skim, and so on.

Microprocessor-based chromatograph systems are now fully established in industry and are continuing to develop in sophistication and complexity. The latest systems are very powerful and flexible, and address the question of reliability and redundancy.

B. Analyzer Computer/Host Computer Data Links

The introduction of microprocessor-based chromatograph systems provided the first opportunity to transmit data from either a single GC or from multiple GCs over a single serial digital data link to the process control computer, (PCC), often called the <u>host</u> computer. Data links of this type are usually interrupt-driven, the connection being through a single serial communication port, and based on either 20-mA current loops or a subset of the RS-232C interface.

Functionality.

Although there are many types of serial data links available from analyzer computer system manufacturers, they all have common functional objectives:

- Checksums, positive acknowledgment, and error recovery procedures are used to ensure integrity and availability of data. The checksum is computed by the analyzer computer system for all messages transmitted, and is equal to the least significant 7 bits of the arithmetic sum of all characters in the message including the "start of header" and the "end of text" characters (defined below). The host computer also computes the checksum correspondingly on all appropriate characters received from the analyzer computer system. This ensures that the message has been transmitted correctly and data are not distorted. Positive acknowledgment by either computer on receipt of a message also ensures that the message has been received correctly. Error recovery procedures ensure that, should no positive acknowledgment be received, a retransmission of the message is initiated.
- Extensive data validation techniques are used to qualify data as acceptable for process control. It is essential that valid data be used by the host computer for process control purposes. Usually, the analyzer computer system will transmit a status code to the host computer which will indicate whether the data are good or bad.
- Compositional data should be presented in engineering units and in a format compatible with wide-dynamic-range data.
- The data link should be able to report not only operating data, but calibration results in terms of response factors. In process chromatography, measured components are calibrated by means of response factors, which relate the area under the component peak to the concentration of that component in the process

stream. These response factors are updated during the calibration. The host computer must be able to recognize that the data transmitted are either process data or calibration data, and act accordingly.

- Both the host computer and the analyzer computer should be able to recognize transmission or link failures.
- The data link should be able to report analyzer status change information to the host computer.
- The question of data link redundancy should be addressed to improve analyzer data availability. The latest analyzer computer systems are designed such that no single component failure in the system will jeopardize the availability of analyzer data to the host computer. For instance, it is often the practice to provide two individual data links to the host computer, with automatic switchover from one to the other in the event of one link failing.
- The hardware requirements for the data link should be as simple as possible.

Communications Protocol

The main difference between the various data links available is in the area of communications protocol and in the extent of the data contained in the particular message transmitted. These functions are usually developed based on specific user requirements and are the reason why so many different protocols exist. However, the protocol is usually based on three distinct functions:

1. Data link establishment to verify either link integrity or I/O buffer availability in the host computer
2. Message transmission, including checksum
3. Error recovery procedure (usually positive acknowledgment or retransmission)

The protocol is therefore "interrupt-driven," where the analyzer computer has control of the data link once the link is established. A typical detailed flowchart of data link implementation is shown in Figure 7. Although the majority of data links are installed as "full duplex," that is, data transmission can be in both directions, they are normally operated in "half-duplex" mode, where data is transmitted only from the analyzer computer to the host computer. However, we are now beginning to see applications where it is necessary to download data from the host computer to the analyzer computer system. For example, in some applications, it is advantageous for the process operator to be able to switch sample streams to the chromatograph from the control room. This involves a download command from the host computer to the analyzer computer.

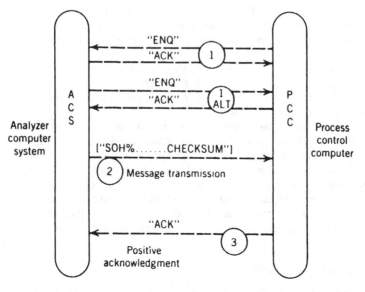

Figure 7 Typical Flow Chart for Data Link Implementation.

Message transmission. Once the data link has been established, the analyzer computer system (ACS) proceeds with transmission of the appropriate message. Messages usually have the following format:

Byte No.	Code	Description
1	"SOH"	Start of header control character
2	Message ID	Message ID character
3 to n − 1	Message text	
n	"ETX"	End of text control character
n + 1	Checksum	Checksum byte calculated by ACS

A typical analysis report is shown in Figure 8. Note that the format provides for the following data to be transmitted:

Analyzer number	Stream number
Analyzer cycle time	Sample inject time
Component concentration	Validity status codes

Byte #	Code	Description
1	"SOH"	Start of header
2	%	Message ID—analysis data
3–4	99	Analyzer number (01–99)
		(in continuous analyzer systems CAC Number)
5–6	99	Stream number (01–99)
7–10	9999	Analyzer cycle time (0001–9999 seconds)
11–15	99:99	Sample inject time (GC) (HR:MN)
16–24	.9999E 9	Component concentration/value
25	9	Validity status code
n	"ETX"	End-of-text
n+1	CHECKSUM	

Figure 8 Typical Analysis Report Format.

Validity status codes. Validity status codes can be presented
in a variety of ways. In the following example, there are four such
codes, with the capability of reporting up to ten:

Status code	Description
0	Calculated component concentration is valid and acceptable for process control.
1	This is a warning condition. Data are still considered valid, but the validation criteria imply impending failure. Analyzer maintenance should be scheduled.
2	Indicates analyzer is temporarily unavailable due to autocalibration.
3	Component value is invalid or unavailable, and the data should be flagged "Bad."

A typical status message is shown in Figure 9.

Disadvantages of the ACS/PCC Data Link

The serial data links described earlier have proved to be an
effective way of transmitting analyzer data to the process control
computer for use in advanced process control schemes. However,
there are some shortcomings, which will be illustrated by reference
to a typical distributed instrumentation and control system, such
as the Honeywell TDC-2000 system with a supervisory process control computer, as shown in Figure 10.

Byte #	Code	Description
1	"SOH"	Start of header
2	#	Message ID—status change
3–4	99	Analyzer number (01–99)
5–6	99	Stream number (01–99)
7	9	Status code
8	"ETX"	End-of-text
9	CHECKSUM	

Figure 9   Typical Status Report Format.

Conventional field transmitters, such as flow, pressure, level and temperature, are hard-wired into the instrumentation system and available for display on the operator's console CRT and to the PCC via the data hiway.   As long as the PCC is running, analyzer data are available to the operator via the computer CRT.   However, should the PCC fail, the analyzer data are no longer available to the operator.   Unfortunately, this is a time when analyzer data would be most useful to the operator in maintaining the process under control in the basic regulatory mode.

Therefore, it has been found essential to provide an alternative path for analyzer data to reach the instrumentation system.   This is usually done by providing analog (4—20 mA DC) output facilities at the analyzer computer system for critical analyzer data, and connecting these outputs directly into the digital instrumentation system via hard-wire cables.   There are two problems with this approach:

1.   An agreement has to be reached with process operations as to what analyzer data are considered critical.   This often results in a conservative approach, calling for extensive backup.
2.   Serious consideration must be given to the cost of purchasing cables, terminations, and input modules for the instrumentation system.

C.   Other Types of Data Links

The chromatograph computer system can be looked upon as a peripheral device from the control system standpoint.   There are many other types of peripheral systems used in process control, examples of which are tank gauging systems, digital temperature indicating systems (DTIs) and, in particular, the programmable logic controller (PLC).   PLCs have been used for some time in industry for relay logic replacement, sequence control, emergency shutdown systems,

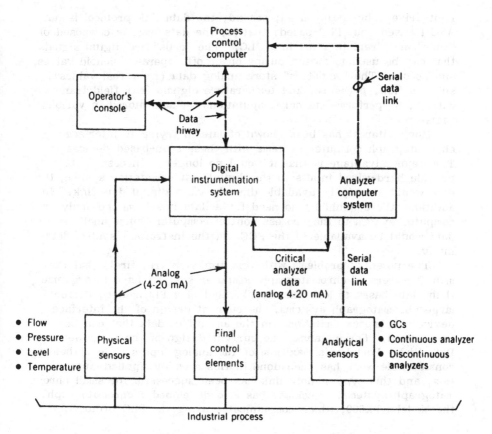

Figure 10 Typical Digital Instrumentation System with Process Control Computer.

and so on. Traditionally, these have been stand-alone devices with no interface, as such, to the process control system.

In recent times, process control systems manufacturers, such as Honeywell, Foxboro, Beckman, and others, have looked seriously at interfacing such peripheral devices with their digital instrumentation systems. The PLC was the first such device to be considered. There are now special devices available that allow the PLC to be interfaced with a digital instrumentation system, examples of which are the Honeywell Data Hiway Port (DHP) and the Foxboro Foxnet Device Gateway (FDG).

The data link between these devices is entirely different from the PCC data link described earlier. First, the link is not inter-

rupt-driven, but scan-based; second, the data link protocol is not
ASCII-based, but PLC-based; that is, the data base is composed of
"coils" and "registers" as in a PLC. The "coils" are digital signals
that can be used to switch pumps on or off, operate solenoid valves,
and so on. The "registers" store analog data (i.e., real values),
such as flow, pressure, and temperature signals from field transmitters, or perform internal computations on data stored in various
registers.

Much interest has been shown of late in trying to interface
chromatograph computer systems with these scan-based devices.
The major advantage is that it would no longer be necessary to
provide hard-wired inputs to the digital instrumentation system; the
data would already be available directly via a digital data link. In
addition, there would be no need for a data link from the analyzer
computer system to the process control computer, since analyzer
data would be available to the PCC via the instrument system data
hiway.

The potential problems with this approach are, first, that the
manufacturers of chromatograph computer systems have to organize
their data bases to look exactly like that of a PLC. Second, for
large chromatograph systems, the current design of the interface
device will impose limitations on the amount of data that can be
transmitted. For instance, the current design of the Honeywell
Data Hiway Port has a maximum of 180 analog inputs. Nevertheless,
considerable work has been done in this area by Applied Automation, and this type of data link has been successful for small chromatograph systems. Beckman has also developed a chromatograph,
the Model CTX600, which communicates using the PLC protocol.

VI. SUMMARY

Gas chromatography has developed over the past 25 years or so into one of the most extensively used on-line analytical techniques in
industrial process control and optimization. Liquid chromatography,
and its several individual techniques, is firmly established in the
laboratory, but its on-line process use has not developed as rapidly
as GC. At the present time, only three companies (Applied Automation Inc., Dionex Corp., and Millipore Corp.) are active in this
area. Nevertheless, substantial growth in on-line process LC is
predicted for the next few years.

The techniques of HPLC (normal-phase and reversed-phase),
IEC, and SEC have great potential in industry as on-line analytical
techniques, including the new field of biotechnology. Computer-
based, multistream, multicomponent systems should find extensive

use in pilot-plant investigations, where their ability to gather large amounts of data (on-line rather than by laboratory testing) could have important implications.

In bioprocess control, undoubtedly the greatest challenge will come in the area of sample-handling technique. On-line chromatography has traditionally involved the sampling and conditioning of fairly conventional process gases and liquids. One exception is in the plastics and elastomers areas, where on-line SEC has been used for polymer MWD measurement. Here the sample is more difficult to handle, and some specialized techniques have been used. In biotechnology, we are treading new ground; nevertheless, it is hoped that some of the experience in sample handling gained in industry over the past 25 years will be of use in this new field.

REFERENCES AND BIBLIOGRAPHY

1. Tswett, M., Ber. Deut. Bot. Gesellsch., 24, 318,384 (1906).
2. Kuhn, R. and Lederer, E., Ber., 64, 1349 (1931).
3. Kuhn, R., Winterstein, A., and Lederer, E., Z. Physiol. Chem., 197, 141 (1931).
4. Adams, B. A. and Holmes, E. L., J. Soc. Chem. Ind., 54, IT (1935); Brit. Pat. 450,308 (1935).
5. Martin, A. J. P. and Synge, R. L. M., Biochem. J. (London), 50, 679 (1941).
6. Eucken, A. and Knick, H., Brennstoff. Chem., 17, 241 (1936).
7. James, A. T. and Martin, A. J. P., Analyst, 77, 915 (1952).
8. James, A. T. and Martin, A. J. P., Biochem. J. (London), 50, 679 (1952).
9. Fuller, E. N., Porter, G. T., and Roof, L. B., J. Chrom. Sci., 20 (1982).
10. Layne, T. G., InTech, Sept. (1984).
11. Mowery, R. A., Jr., in Proc. ISA Conf., Houston, TX (1980).
12. Mowery, R. A., Jr., Chem. Eng., May (1981).
13. Snyder, R. L. and Kirkland, J. J., Introduction to Modern Liquid Chromatography, 2nd ed., Wiley, New York (1974).
14. Clevett, K. J., Process Analyzer Technology, Wiley, New York (1986).

5

Applications of NADH-Dependent Fluorescence Sensors for Monitoring and Controlling Bioprocesses

JOHN H. T. LUONG and ASHOK MULCHANDANI   *National Research Council of Canada, Montreal, Quebec, Canada*

I. INTRODUCTION

Bioconversion processes have been exploited throughout the ages to manufacture useful products; beer, wine, amino acids, and antibiotics are obvious examples. Since the productivity of bioprocesses can be enhanced by maintaining the appropriate biological and environmental conditions, microbial process control can be implemented to optimize the performance of the bioreactors and their satellite hardware.

The first aspect of control is the ability to measure the current state of the process and compare it with the desired state in order to initiate actions to maintain control. Therefore, principal in the development of such process control practices is the provision of suitable sensors.

Instrumentation systems can be classified into three basic categories: off-line, on-line, and in-line. Off-line techniques require the removal of samples for subsequent analysis by wet-chemical methods or automated laboratory instrumentation systems. Such techniques are time-consuming, cumbersome, and inapplicable for any kind of on-line monitoring and feedback control scheme. On-line systems utilize the continuous sampling of a process stream with the subsequent analysis such that the response time is well within the range of meaningful process control decisions. A classical example of such systems includes sampling of gas streams for on-line analysis by gas chromatograph. In-line systems are those where sensors are directly in contact with the liquid broth and provide rapid response signal. Unfortunately, at the present time, such measurements are limited to physical and chemical quantities such as temperature, pressure, aeration, agitation, viscosity, pH, redox, dissolved oxygen, and carbon dioxide. The most important biological parameters, such as biomass, substrates, products, metabolites, and enzyme activities, are still not accessible to either on-line or in-line measurements.

The development of sensors for monitoring and controlling bioreactors is hampered by several factors. Besides good response time, accuracy, sensitivity and stability, the devices must be noninvasive and capable of sterile operation. In addition, many food and pharmaceutical processes demand a steam sterilization procedure to prevent microbial contamination. Clearly, this is incompatible with several sensors unless one has a separate bleed-off from the main process to a remote testing site.

Many biological compound such as proteins, enzymes, coenzymes, pigments, and primary or secondary metabolites emit characteristic fluorescence light after excitation by light from the visible or near-UV region (Table 1). Among these compounds, reduced nicotinamide adenine dinucleotide (NADH) and reduced nicotinamide adenine dinucleotide phosphate (NADPH) are the most important, since they are present in all living cells and play an important role in the microbial energy metabolism. NADH-dependent fluorescence can therefore be used to monitor biological compounds.

This chapter focuses on the development and application of NADH-based sensors for quantifying the cell concentration and monitoring their metabolic state and for determining specific compounds that could be of interest in a biological process. Use of NADH-dependent sensors in control of bioprocesses will also be discussed.

II. PRINCIPLE OF NADH-FLUORESCENCE MEASUREMENT

Fluorescence is the emission of light from a molecule after it has absorbed radiation with a spectral shift toward longer wavelength. Generally, fluorescence emission occurs very rapidly after excitation (10^{-6} to 10^{-9} s), and it cannot be perceived by the naked eye. Molecular emission is a particularly important analytical technique because of its extreme sensitivity and good specificity. This high sensitivity is due to the fact that the signal is measured over a zero background in comparison with the spectrophotometric method, where the signal measured is the difference between the incident and transmitted light. In the latter, the small fluctuations of signal are lost in the background signal with a corresponding large decrease in the sensitivity. Noteworthy is the fact that specific chemical compounds possess specific excitation-emission patterns, allowing

Table 1 Examples of Biologically Important Fluorescent Substances

	Excitation peak (nm)	Fluorescence peak (nm)
Tyrosine[a]	275	303
3,4-Dihydroxyphenylalanine (Dopa)	345	410
Tryptophan[a]	287	348
Kynurenine	370	490
5-Hydroxytryptamine (Serotonin)	295	330
Phenylalanine	260	282
3,4-Dihydroxyphenylethylamine (Dopamine)	345	410
Histamine	340	480
Vitamin A	372	510
Flavins	450	535
NADH & NADPH	340	460
p-Aminobenzoic acid	294	345
Vitamin B-12	275	305
Estrogens	285	325
ATP, ADP, adenine, adenosine	272	380

[a]Responsible for protein fluorescence.

detection of such a compound in a mixture of chemicals. Hence, it would be possible to monitor a particular interesting flourescing compound using its specific excitation-emission pattern. Many aromatic, heterocyclic, and polycyclic compounds are fluorescent. Compounds with multiple conjugated double bonds on electron donating groups such as $-OH$, $-NH_2$, and $-OCH_3$ are favorable to fluorescence. Groups such as $-NO_2$, $-COOH$, $-CH_2COOH$, $-Br$, $-I$, and azo tend to inhibit fluorescence [1].

The instrumentation of fluorescence measurement in aqueous solutions is very similar to that used in spectrophotometry, differing only in a right-angle measurement rather than a straight-line

one and in the use of a second filter on the monochromator [2].
In fact, almost all commercial spectrophotometers, such as the Beckman DU or DK, can be easily converted to a fluorometer. A UV source such as a mercury vapor lamp is sufficient for many applications since most fluorescing molecules absorb UV radiation over a band of wavelengths. The primary filter is used to filter out the wavelength close to the wavelengths of the emission because, in practice, some radiation is scattered. The second filter passes the wavelength of emission but not the wavelength of excitation. A spectrofluorometer, instead of filters, incorporates two monochromators, one to select the wavelength of excitation and one to select the wavelength of fluorescence.

NADH-dependent fluorescence can also be measured in whole cells (henceforth referred to as "culture fluorescence"). The principle is applied to any cell type whether bacteria, fungi, plant, or animal. However, the specific amount of NADH per cell appears to vary with a great range. Harrison and Chance [3] were the first to build an instrument capable of measuring culture fluorescence within a fermenter. Since then, the applications of culture fluorescence for monitoring and controlling bioprocesses have gained considerable attention. Two different types of fluorometers, designed for the in situ monitoring of microbial cells have appeared on the market (BioChem Technology Malvern, PA, USA, and Dr. W. Ingold, AG , Urdorf, Switzerland). The probe is steam sterilizable and can be applied to monitor the fluorescence of a suspended cell population in the vicinity of the sensor. The sensor excites the fermentation liquid with a low-intensity, short-wavelength (340-nm), UV lamp. The process liquid in turn emits light energy in the visible spectrum (460 nm) (Figure 1).

The Biochem Technology or Ingold fluorosensor comprises a sensor unit and an associated electronic unit. The sensor unit is compatible with industry standard 25-mm-diameter Ingold fittings for easy insertion in bioreactors. The optical head of the sensor unit is completely immersed in the liquid when mounted. It can be subjected without any problem to the normal in situ sterilization procedure, since all parts coming into contact with the sample medium are heat-resistant up to a temperature of 130°C. After sterilization, the sensor unit must be in thermal equilibrium with its surroundings (cooling-off time about 1/2 h). The measuring temperature range is 0−80°C with a maximum permissible pressure of 6 bar.

The electronic unit nests in a standard 19-in relay rack or bench enclosure. It can operate on 115 V, 60 Hz or 230 V, 60 Hz and provides a 4−20 mA output, which allows the signal to be transmitted more than a few feet without losing accuracy and fidelity. It also houses the lamp power supply and test circuitry. Ei-

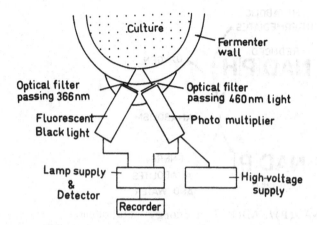

Figure 1   Schematic design of the in-line culture fluorescence probe.

ther the Biochem or Ingold fluorosensor has been designed for continuous operations, and requires very little maintenance.   The
equipment is supplied tested and ready for use.
        Calibration of the fluorosensor can be performed using a suitable
fluorescent solution.   NADH is not very suitable for this purpose
since it is relative unstable in solution and very expensive.   Quinine and thiamine S with fluorescence characteristics similar to NADH
can be used to check the functioning of the fluorosensor.   Since
such compounds are sensitive to light and oxygen, a freshly prepared stock solution should be used for each assay.

III.   MEASUREMENT OF CULTURE FLUORESCENCE

The coenzymes NAD(P)/NAD(P)H serve as an electron carrier in
various intracellular redox reactions.   During aerobic metabolism,
NADH serves as an electron donor to the enzymes of the respiratory
chain and is oxidized with formation of water.   In an anaerobic metabolism, NAD(P)H produced in glycolysis donates the electron to
reducible intermediates such as acetaldehyde, butyraldehyde, ethanol, etc. (Figure 2).
        Duysens and Amesz [4] first observed in 1957 that the excita-
tion of whole cells of Saccharomyces cerevisiae (baker's yeast) by
UV light led to a fluorescent light at about 450 nm.   The fluorescence spectrum of whole cells was similar to the spectrum of NADH
in solution (Figure 3).   About 20 years later, Zabriskie and
Humphrey [5] also presented evidence that the fluorescence observed
with whole cells of bakers' yeast is based on the cellular NADH content.

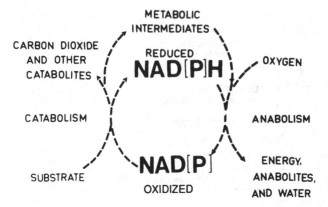

Figure 2   Role of NAD(P)/NAD(P)H in energy metabolism.

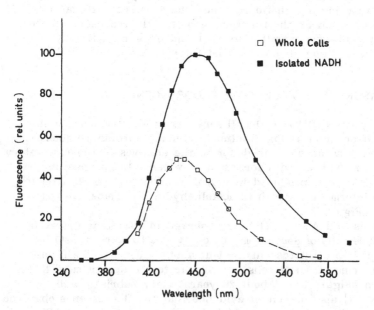

Figure 3   Fluorescence spectra of isolated NADH and of whole cells.

The culture fluorescence is a composite signal since it is affected by several parameters, such as population of viable cells, metabolic state of the cells, and environmental effects (pH, temperature, etc.). Quantitatively, the culture fluorescence is expressed by the following equation [6]:

$$NFU(t) = Y_{f/x} [1 + M(t)] X + E(t) + P_r(t)$$

where

$Y_{f/x}$ = fluorescence yield per cell in its lowest reducing state

$M(t)$ = component of the fluorescence yield due to metabolic changes

$E(t)$ = extracellular environmental fluorescence

$P_r(t)$ = noncellular process concentration effects

X = biomass concentration

One NFU corresponds to a change in fluorescence caused by 0.122 μM NADH at 30°C and pH 8 in the concentration range of 1.0 to 25.0 μM NADH [6].

In practice, aeration and agitation are usually maintained constant throughout the course of fermentation. Therefore, the primary information obtained through fluorescence of the culture is only related to cell concentration and microbial metabolic state. Under a given metabolic state, the NAD(P)H pool per unit biomass is relatively constant. This condition is applicable when the cells are grown in continuous culture or during the exponential growth phase of batch culture. Under such a condition, the culture fluorescence is only dependent upon the cell concentration. This expectation was experimentally confirmed by Zabriskie and Humphrey [5], who proposed a nonlinear equation for correlating the fluorescence data with the cell concentration (Figure 4). Since then, in-line fluorescent measurement has been applied to several bioconversion processes employing different microorganisms such as bacteria, yeast, fungi, mammalian cells, plant cells, and hybridoma cells (Table 2) [7–19]. In many cases, fluorescence data were proven to be a relative indicator of the biomass concentration. However, a quantitation in terms of moles of NADH per cell is yet to be established. From a process monitoring perspective, in-line measurement of biomass concentration is of advantage since it enables one to monitor continuously the progress of a fermentation. Such information provides a means of comparing the performance of a particular batch with earlier batches and an opportunity to take any designated corrective action, if necessary. Following the biomass concentration in the bioreactor reliably and without time delay has always been a

Table 2 Organisms Used for Fluorescence Measurement

Bacteria	Fungi	Mammalian cells	Reference
Escherichia coli			[6,7]
Clostridium acetobutylicum			[8]
Methylomonas mucosa			[9]
Alcaligenes eutrophus			[10]
Pediococcus sp.			[11]
Zymomonas mobilis			[12]
Pseudomonas putida			[13]
Bacillus subtilis			[7]
	Saccharomyces cerevisiae		[4,5,14]
	Penicillum chrysogenum		[12]
	Streptomyces sp.		[4]
	Candida sp.		[15,16]
	Trichosporon cutaneum		[17]
		Mouse fibroblast	[11]
		Melanoma	[18]
		Lymphocytes	[11]
		Hybridoma	[19]

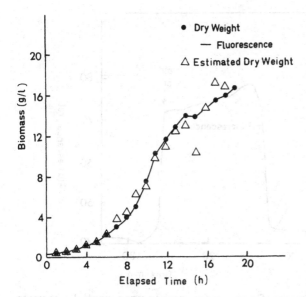

Figure 4 Comparison of biomass concentration determined analytically and estimated from the fluorescence signal.

bothersome task, often representing a major problem, particularly with fermentation broths containing suspended solids other than microbial cells.

The NADH-dependent fluorescence information can be used for monitoring changes from aerobic to anaerobic metabolism [15]. As mentioned previously, NADH serves as an electron donor to the enzymes of the respiratory chain. Therefore, NADH increases when the cultivation has limited oxygen and decreases when oxygen is in excess (Figure 5) [15]. This phenomenon has been demonstrated for a variety of species, such as baker's yeast, Trichosporon cutaneum, Klebsiella aerogenes, and Candida tropicalis. Obviously, such information can be exploited to provide a very accurate dissolved oxygen control based on the intracellular state of the cell rather than on the dissolved oxygen level in the liquid broth. This method of controlling dissolved oxygen will be more energy efficient because it will avoid the usual practice of overagitation and overaeration and provide superior control due to rapid response.

The NADH-dependent culture fluorescence measurement has been applied to develop process control strategies. An interesting example of such applications was the work initiated by the researchers of BioChem Technology of Malvern, PA [11]. The cellular NADH, and therefore NADH-dependent flourescence, in baker's

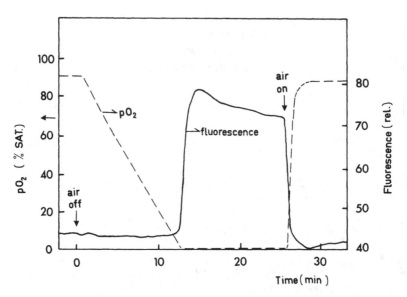

Figure 5   NADH-dependent fluorescence and oxygen partial pressure during aerobic-anaerobic transitions of <u>Candida</u> <u>tropicalis</u>.

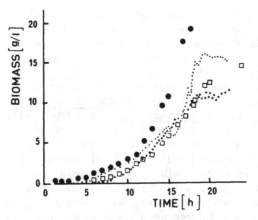

Figure 6   Comparison of the performance of baker's yeast process using NADH-based and respiratory quotient-based substrate feeding control on glucose strategy.   (□) fluorescence of resting cells; (●) fluorescence of cells growing anaerobically; (⁺ ⁺ ⁺ ⁺) RQ-based control; (· · · ·) fluorescence-based control.

yeast cell is a function of the cultivation condition. It is maximum when the cells are metabolizing glucose in an anaerobic state, followed by aerobic growth on ethanol, and minimum when the cells are in a resting state. On the basis of such information, a glucose-feeding strategy was developed so that the culture fluorescence was controlled between the two metabolic limits: anaerobic growth on glucose and resting cells. The feed pump was activated when the measured fluorescence approached the simulated resting cell fluorescence and was shut off when it reached the upper limit of anaerobic growth on glucose. This methodology was reported to be comparable to the conventional respiratory quotient (RQ) control algorithm (Figure 6).

In batch processes employing recombinant cells the product yield per cycle can be maximized by maximizing the biomass yield per liter of the culture before induction for product expression. Furthermore, since induction imposes a severe metabolic drain on the cells, the timing of induction is critical. Conventionally the cell density is followed off-line turbidometrically. This method is laborious and subject to variation from batch to batch. However, in-line fluorescence measurement which provides real time biomass concentration data is more efficient. A control action can then be automatically triggered to induce product synthesis after the desired cell density is achieved. Brandis successfully employed this control action to induce product synthesis in temperature-inducible strains of Escherichia coli [20].

The fast response of the NADH-based fluorescence monitoring can also be combined with other slow responding signals to set up an integrated control strategy. This was demonstrated by Meyer and Beyeler [21] for substrate feeding strategy in continuous production of baker's yeast. The process input substrate feed rate (or dilution rate) was controlled by two controllers linked together (Figure 7). The fluorescence controller reacted to the fast process changes according to proportional-differential (PD) characteristics. A second controller, which reacted to the carbon dioxide production rate (CPR) (calculated on-line from effluent gas analysis data) responded to the long-term process variation. The results of the process controlled this way (Figure 7) showed that at low substrate feed rate the CO_2 controller always tried to increase the substrate feed rate to the set point. However this increase was limited by the fluorescence controller when the fluorescence signal responded too fast. Doing this, the metabolic change was prevented and the control goal of high productivity was achieved within reasonable time.

Other applications of culture fluorescence are also noted. Mixing time behavior in bioreactors during cultivation can be evaluated by using the fluorescence approach [22]. Mixing times can be

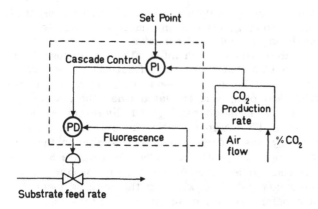

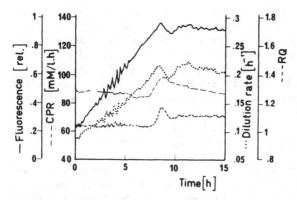

Figure 7   Control of baker's yeast fermentation using flourescence
and carbon dioxide production rate controllers.

studied by either following a fluorescent tracer (quinine) or by
monitoring the change in NADH-dependent fluorescence resulting
from added substrate.   Mixing time determined by the latter method
includes all physical and biochemical steps necessary until a substrate reaches the final place within the cell, whereas the former
way indicates only the physical and, consequently, the equipment-
dependent part.

Culture fluorescence measurements can also be utilized to study
substrate transport into microorganisms.   This approach was initiated by Einsele et al. [23], who followed intracellular NADH content
of baker's yeast and Candida tropicalis in response to step-changes
in substrate concentration.   Of course, all other parameters such as
pO_2, pH, and temperature were kept constant during the measurement.

Fluorescence is more sensitive than spectrophotometric measurements (10^{-12} $\underline{M}$ versus 10^{-6} $\underline{M}$). The determination of culture fluorescence, however, was reported to be less sensitive than calorimetry, a technique used to follow the heat of reaction [13]. In general, fluorescence techniques probably are less time consuming than chromatography or mass spectrometry. This is a definite advantage both in continuous process control and when large numbers of samples have to be assayed.

Fluorescence techniques, however, possess several shortcomings. One of the most serious problems associated with the use of fluorescence is the "inner-filter" effect on fluorescent quenching. Highly absorbing molecules (i.e., dichromate) rob energy from the molecule under study, lowering the total fluorescence observed. Other interferences are due to molecules that absorb and fluoresce at the same wavelengths as the substance being determined. Proteins, for instance, cause serious interference in fluorescence measurements made in the UV region because they contain amino acids such as tryptophan and tyrosine that are fluorescent in this region [24]. The inner-filter effect can also arise from too high a concentration of fluorophore itself. Some of the analyte molecules will reabsorb the emitted radiation of others. Mathematical models were recently developed to correct for interference due to inner-filter effects in the measurement of culture fluorescence [25]. The validity of such models was proven with well-defined synthetic systems. The models were also capable of providing a conceptual framework for correlating the measured fluorescent intensity to cell density and metabolic state of the cell (baker's yeast).

The culture fluorescence measurement also requires a good control of the operating conditions such as pH, temperature, and oxygen (if the process is aerobic). It is always a difficult task to maintain such control for highly viscous fermentation broths (polysaccharides production such as xanthan, pullulan, dextran, etc.). Furthermore, when the fermentation medium is dark in color due to the presence of molasses, corn steep liquor, or intracellular pigments, as in the case of pullulan production by Aureobasidium pullulans [26], the physical parameters of light absorption and emission are adversely affected and fluorescence may not be suitable.

IV. DEVELOPMENT OF NAD(P)H-BASED BIOSENSORS

In addition to biomass, efficient control of biological process requires the measurement of substrate level, product formation, and metabolites. The principle of NADH-dependent fluorescence can be extended to all NAD- and NADP-dependent reactions involved in enzymatic analysis:

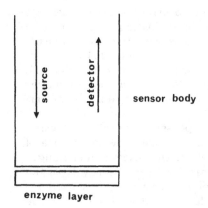

enzyme layer

Figure 8 Schematic of NAD(P)H-based biosensor.

$$\text{Substrate + NAD} \xrightleftharpoons[]{\text{dehydrogenase}} \text{Product + NADH}$$

A wide range of compounds can be assayed by NADH-based biosensors, since several NAD-NADP-dependent enzymic reactions have
been known (Table 3). Specific compounds such as glucose, ethanol, lactate, xylitol, glycerol, etc. could be of interest for a biological process.

Conceptually, the dehydrogenase enzyme can be immobilized on
the tip of an optical device to form a novel class of biosensor. The
analyte diffuses to the immobilized biocatalytic layer, and NADH is
either consumed or produced at the sensor tip (Figure 8). The
development of NADH-based biosensor was pioneered by Luebbers
and Opitz [27], who immobilized lactate dehydrogenase on a fiber
tip. Consequently, interaction between lactate and lactate dehydrogenase produced NADH which was then monitored fluorometrical-
ly. This approach was also attempted for the determination of ethanol via alcohol dehydrogenase.

The NADH-based biosensor is not expected for in-line measurement in bioreactors since the enzyme itself cannot withstand steam
sterilization and the probe responds to the cell's mass as well. In
this case, fermentation broth must be filtered prior to analysis. On-
line application of the NADH-based biosensor, however, is feasible
by forming a continuous circulating loop with the fermentation vessel. As shown in Figure 9, liquid is pumped with a peristaltic
pump from the fermentation vessel and through the radial bypass
filter (ceramic or stainless steel). Broth not filtered returns to the
vessel, thus minimizing the loss of liquid inside the bioreactor. The

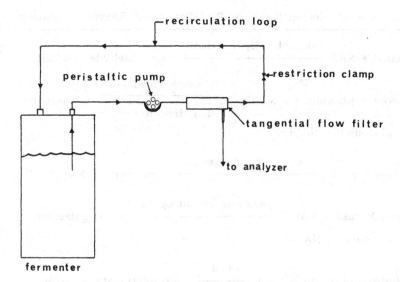

Figure 9   Schematic for on-line measurement of fermentation compounds.

cell-free liquid is then delivered to a detection chamber equipped with a NADH-based biosensor.

   The NADH-based biosensor is extremely sensitive, and determinations at the ppb level are common.   The high sensitivity presents some problems since organic substances of the liquid culture usually possess concentrations of several orders of magnitude higher than the above-mentioned level.   This shortcoming can be easily overcome by adding a dilution stream to the analyte stream.

   An innovative approach to NADH-based biosensors is being developed where the enzyme is located inside the sensor body (Figure 10) [28].   The enzyme and all other reagents required for the analytical reaction are placed in the internal solution.   A gas-permeable membrane separates this internal solution from the sample.   This device is restricted to volatile analytes since the substrate must diffuse through the membrane.   A high degree of selectivity is expected since both a membrane barrier and a selective enzyme are used together in the sensor operation.   The enzyme activity is not adversely affected by the sample components, and optimal conditions can be established for the analytical reaction without altering the sample.

   It should be remembered that fluorescence is a photochemical reaction appearing at the molecular level.   The measurement of fluorescence is thus very sensitive to the immediate environment.

Table 3 A Few Selected NADH-NADPH-Dependent Enzymic Reactions

---

- Ethanol + NAD $\xrightarrow{\text{alcohol dehydrogenase}}$ acetaldehyde + NADH

- Glucose-6-phosphate + NADP $\underset{\text{dehydrogenase}}{\overset{\text{glucose-6-phosphate}}{\rightleftharpoons}}$ 6-phospho-

    gluconate + NADPH

- Lactate + NAD $\xrightarrow{\text{lactate dehydrogenase}}$ pyruvate + NADH + H$^+$

- β-D-Galactose + NAD $\xrightarrow{\text{galactose dehydrogenase}}$ D-galactono-

    γ-lactone + NADH + H$^+$

- α-Hydroxybutyrate + NAD $\underset{\text{HBDH}}{\overset{\text{pH 9}}{\longrightarrow}}$ α-ketobutyrate + NADH

    (HBDH = α-hydroxybutyrate dehydrogenase)

- Indole acetaldehyde + NAD $\underset{\text{dehydrogenase}}{\overset{\text{aldehyde}}{\rightleftharpoons}}$ indoleacetic acid

    + NADH

- D-Sorbitol + NAD $\underset{\text{dehydrogenase}}{\overset{\text{sorbitol}}{\rightleftharpoons}}$ D-fructose + NADH + H$^+$

- Acetaldehyde + $\underset{\text{dehydrogenase}}{\overset{\text{aldehyde}}{\longrightarrow}}$ acetate + NADH + H$^+$

- Malate + NAD $\underset{\text{dehydrogenase}}{\overset{\text{Malate}}{\rightleftharpoons}}$ oxaloacetate + NADH + H$^+$

- L-Alanine + NAD $\underset{\text{pH 8.8–9}}{\overset{\text{pH 10–10.5}}{\rightleftharpoons}}$ pyruvate + NH$_3$ + NADH

Table 3 (Continued)

- D-Isocitrate + NADP $\xrightarrow{\text{D-isocitrate dehydrogenase}}$ α-oxoglutatate + CO_2

  + NADPH + H^+

- β-D-Glucose + NAD(P) $\underset{\text{dehydrogenase}}{\overset{\text{glucose}}{\rightleftarrows}}$ D-glucono-δ-lactone

  + NAD(P)H

- L-Glutamate + NAD $\underset{\text{dehydrogenase}}{\overset{\text{glutamate}}{\rightleftarrows}}$ 2-oxoglutarate + NADH

  + NH_4^+

- Glycerol + NAD $\underset{\text{dehydrogenase}}{\overset{\text{glycerol}}{\rightleftarrows}}$ dihydroxyacetone + NADH

  + H^+

- Myoinositol + NAD $\underset{\text{dehydrogenase}}{\overset{\text{inositol}}{\rightleftarrows}}$ 2,4,6/3,5-pentahydroxy

  cyclohexanone + NADH + H^+

- Xylitol + NAD $\xrightarrow{\text{sorbitol dehydrogenase}}$ D-xylulose + NADH + H^+

- D-Manitol + NAD $\xrightarrow{\text{Manitol dehydrogenase}}$ fructose + NADH + H^+

- 3-α-Hydroxysteroid + NAD $\xrightarrow{\text{hydroxysteroid dehydrogenase}}$ 3-ketosteroid

  + NADH

---

optical fibers

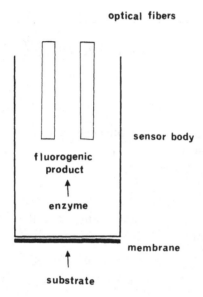

sensor body

fluorogenic
product

↑

enzyme

membrane

↑

substrate

Figure 10   Novel NADH-based biosensor design.

    The fluorescence emitted by a molecule can change with temper-
ature, dissolved oxygen, pH, viscosity, gas bubbles, ionic strength,
and chemical composition of the surrounding medium. It was reported that an increase in temperature of 1°C can reduce the fluorescence of tryptophan by about 5% [2]. Hence, a vigorous control
of the temperature must be made during the measurement. The
author also reported that dissolved oxygen can be a severe inhibitor of the fluorescence, and a typical reduction of 20% can be observed with a dissolved oxygen concentration as low as 10^{-3} $\underline{M}$.
The control of pH is also very important and is thought to affect
the fluorescence by changing the ionic state of the fluorophor as
well as its immediate environment [5]. The increase in the apparent
viscosity of the medium will also increase the fluorescence of compounds.

V.   CONCLUSIONS

Monitoring by fluorescence optical sensors allows a continuous and
remote control of bioconversion processes. Besides biomass, several
interesting compounds such as ethanol, glucose, glycerol, etc., can
be detected. Fluorescence-based sensors possess certain advan-

tages and special features in comparison to commonly used electrochemical sensors. The optical sensor requires no reference, and its sensitivity can be tuned to the desired range of measurement, resulting in a higher resolution. In contrast to electrodes, the optical sensor does not require isolation and shielding of electrical circuits. Notwithstanding these advantages, interference by ambient light is a potential source of error. This disadvantage can be overcome by either complete optical isolation of the sensor head or by light modulation.

Culture fluorescence has been proven as an index of biomass concentration even though a quantitation in terms of fluorescence per cell is yet to be established. On-line estimation of the viable biomass concentration and control of substrate and oxygen supply in the study of substrate uptake kinetics and regulation are just two important examples of the immense field of possible applications of this method. Similar to enzyme electrodes, fluorescence enzymatic sensors can be developed by combining the optical element with a layer of immobilized dehydrogenase enzymes. Conceptually, a wide range of interesting subtances involved in NAD-NAD(P)-dependent reactions can be monitored by such devices. Undoubtedly, NADH-based biosensors are qualified as an alternative tool for determination of various biological and chemical substances. The future developments of optical components such as semiconductor detectors or fiber optics will certainly increase the versatility of this technique.

REFERENCES

1. Christian, G. D., Analytical Chemistry, 3rd ed., Wiley, New York (1980).
2. Guilbault, G. G., Practical Fluorescence, Theory, Instrumentation and Practice, Marcel Dekker, New York (1967).
3. Harrison, D. E. F. and Chance, B., Appl. Microbiol., 19, 446 (1970).
4. Duysens, L. N. M. and Amesz, J., Biochim. Biophys. Acta., 24, 19 (1957).
5. Zabriskie, D. W. and Humphrey, A. E., Appl. Eur. Microbiol., 35, 336 (1978).
6. Armiger, W. B., Zabriskie, D. W., Maenner, G. F., and Forro, J. F., Proceedings of Biotech 84, 2, 601 (1984).
7. Meyer, H. P., Beyeler, W., and Fiechter, A., J. Biotechnol., 1, 341 (1984).
8. Srinivas, S. P. and Mutharasan, R., Biotechnol. Lett., 9, 139 (1987).

9. Luong, J. H. T. and Carrier, D. J., Appl. Microbiol. Biotechnol., 24, 65 (1986).
10. Groom, C. A., Luong, J. H. T., and Mulchandani, A., J. Biotechnol., 8, 271 (1988).
11. The FluroMeasure System, User's Manual, BioChem Technology, Inc., Malvern, PA (1987).
12. Scheper, Th., Lorenz, Th., Schmidt, W., and Schügerl, K., J. Biotechnol., 3, 231 (1986).
13. Samson, R., Beaumier, D. and Beaulieu, C., J. Biotechnol., 6, 175 (1987).
14. Scheper, Th. and Schrügel, K., Appl. Microbiol. Biotechnol., 23, 440 (1986).
15. Beyeler, W., Einsele, A., and Fiechter, A., Eur. J. Appl. Microbiol. Biotechnol., 13, 110 (1981).
16. Ristroph, D. L., Watteeuw, C. M., Armiger, W. B., and Humphrey, A. E., J. Ferment. Technol., 55, 559 (1977).
17. Gschwend, K., Beyeler, W., and Fiechter, A., Biotechnol. Bioeng., 25, 2789 (1983).
18. Leist, C., Meyer, H. P., and Fiechter, A., J. Biotechnol., 4, 235 (1986).
19. Armiger, W. B., Forro, J. R., Lee, J. F., MacMichael, G., and Mutharasan, R., Proc. Bio '86, London, England (1986).
20. Brandis, J., BioChem Technology Update, Vol. II, No. III, BioChem Technology, Malvern, PA (1987).
21. Meyer, C. and Beyeler, W., Biotechnol. Bioeng., 26, 916 (1984).
22. Einsele, A., Ristroph, D. L., and Humphrey, A., Biotechnol. Bioeng., 20, 1487 (1978).
23. Einsele, A., Ristroph, D. L., and Humphrey, A., Eur. J. Appl. Microbiol. Biotechnol., 6, 335 (1979).
24. Guilbault, G. G., Handbook of Enzymatic Methods of Analysis, Marcel Dekker, New York (1976).
25. Srinivas, S. P. and Mutharasan, R., Biotechnol. Bioeng., 30, 769 (1987).
26. Luong, J. H. T. and Mulchandani, A., unpublished results.
27. Luebbers, D. W. and Opitz, N., Sensors and Actuators, 4, 641 (1984).
28. Arnold, M. A., in Biosensors International Workshop 1987 (R. D. Schmid, G. G. Guilbault, Z. Karube, H.-L. Schmidt, and L. B. Wingard, eds.), GBF Monographs, 10, Braunschweig (1987), pp. 223–237.

6

Fiber-Optic Sensors in Bioprocess Control

OTTO S. WOLFBEIS *Institute of Organic Chemistry, Karl Franzens University, Graz, Austria*

I. INTRODUCTION

Chemical sensing with optical fibers is considered to be one of the most promising of the emerging sensor technologies. Various reviews on this subject have appeared [1–11], but only a few are specifically written for the scientist interested in the field of biotechnology. Doubtless, there is a considerable demand for sensors in this area, and various electrochemical devices have been used more or less successfully in the last few years. Optical methods such as absorptiometry and fluorimetry, on the other hand, have been used for long periods and are more or less established in the analytical laboratory, but so far have been confined to off-line methods rather than for continuous sensing. A major breakthrough has been achieved only in the past 10 years when optical spectrometry was coupled to immobilization and fiber-optic technologies. This combination allows the construction of the optical sensor together with the transportation of optical information over long distances from the sample to the meter. Fibers are robust and can therefore be exposed to varying, even hostile, conditions, whereas the rather sensitive optical instrument can remain in a central laboratory.

Depending on the field of application, fiber-optic chemical sensors (FOCS) can offer advantages over other sensor types that make them particularly useful in biotechnology sciences. The following are considered to be most important:

1. Most optical sensors do not require a reference signal such as is necessary in all potentiometric methods where the difference of two absolute potentials is measured.
2. The ease of miniaturization allows the development of very small, light, and flexible instrumentation useful for remote sensing over distances of typically 2 to 100 m.

3. Because the primary signal is optical, it is not subject to electrical interferences by, e.g., static electricity of reaction vessels or surface potentials of the sensor head. Optical fibers
   also bestow electrical isolation at the transducing site (usually
   the fiber tip), thus making the device safe for use in explosive
   environment (such as in methane/air mixtures) and in in vivo
   applications.
4. Analyses can be performed in almost real time, and no sampling
   with its inherent drawbacks is necessary.
5. Coupling of small sensors for different analytes to produce a
   bundle of small size allows simultaneous monitoring of various
   analytes, such as pH, oxygen, and carbon dioxide, by hybrid
   sensors without crosstalk of the single strands.
6. Optical waveguides allow a high flux of information density to
   be maintained. Thus, red and green light may be sent through
   the same fiber into one direction, and yellow and blue light into
   the other direction simultaneously.
7. Many fiber-optic sensors can be heat sterilized.
8. FOCS can offer cost advantages over other sensor types. Most
   fibers are made from either plastic or glass, both of which are
   inexpensive. It is therefore likely that a significant part of
   optical sensors can be designed for single use.

Notwithstanding these advantages over other sensor types, fiber
sensors can exhibit the following disadvantages:

1. Ambient light interferes. Consequently, the sensor has to be
   applied in a dark environment, or must be optically isolated, or
   its signal must be encoded to be able to differentiate it from
   background light. Many samples have an intrinsic fluorescence
   that may interfere, and this can only be compensated for by
   proper isolation of the optical system of the fiber sensor from
   the optical properties of the sample.
2. Sensors having immobilized indicator phases are likely to have
   limited long-term stability because of photobleaching or washout.
3. Sensors with immobilized pH indicators or chelating reagents
   have limited dynamic ranges as compared to electrodes which
   have linearity over a wide concentration range. The respective
   association equilibria obey the mass action law, and the corresponding plots of optical signal versus log of analyte concentration are sigmoidal rather than linear as in the case of the Nernst
   relation.
4. Commercial accessories of the optical system are not optimal yet.
   Stable and long-lived light sources, less expensive lasers, more
   intense blue LEDs, better connectors, terminals, and optical fi-

bers are needed. The lack of violet LEDs and inexpensive blue
or violet lasers is particularly annoying.

Among the potential fields of application for FOCS, mention
should be made of remote sensing of environmental parameters in
air, seawater or groundwater, continuous monitoring of chemical parameters in chemical plants and nuclear power stations, and in de-
fense. Since many plastic fibers can be made biocompatible, they
may be applied to invasive methods in medicine. It is widely accepted now that biotechnology provides one of the largest fields of
chemical sensor application. This chapter gives an overview on the
principles and applications of optical sensors, with particular emphasis given to problems associated with bioprocess control.

II. OPTICAL WAVEGUIDES

Fibers are an outgrowth of the communication industry. Suggested
as a means for long-distance communications in 1968, they are now
gradually replacing classical electric cables because they are cheaper
than metal wires and now have signal attenuations as low as 1 dB
or less per kilometer at 1.5-μm wavelength. Essentially, a fiber
consists of an optically transparent material called the core, which
guides the light, and a cladding, which covers or surrounds the
core. A schematic of the path of light in a fiber is shown in Figure
1. Light is totally reflected at the clad/core interface and therefore can be transported without significant loss over fairly large
distances.

In a typical arrangement for performing remote analysis (Figure
2), light is focused into the core and guided to a sample to produce

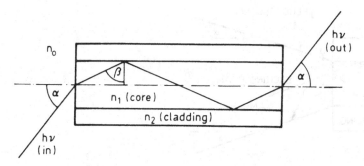

Figure 1   Path of light inside an optical waveguide.

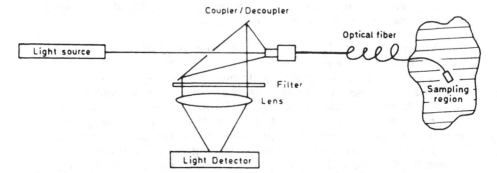

Figure 2   Optical arrangement for performing fiber-optic remote analyses.

analytical information. The optical signal (e.g., fluorescence) returns through the same (or another) fiber, and its intensity is measured. The input beam may be considered as the question, and the returning light as the encoded answer. Similar instrumentation may be used when absorbance, reflectance, or Raman scatter intensity are to be measured. A schematic of the path of light in a fiber tip as used for performing remote fluorimetry is given in Figure 3.

The basic physical principles underlying total reflection within an optical waveguide are simple: Light is totally reflected at an interface between two optical phases when the index of refraction of the outer phase is smaller than the index of refraction of the core. The relation between the two indices of refraction is given by

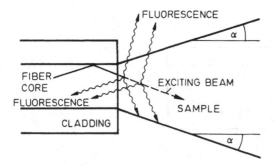

Figure 3   Schematic of fluorescence produced at a single fiber termination.

$$\frac{n_1}{n_2} = \sin \beta \tag{6.1}$$

where β is the limiting angle of reflection (see Figure 1). Another useful parameter for the spectroscopist is the angle within which light can be launched into a fiber. This angle (2α) depends on the refractive indices of core (n_1) and cladding (n_2) and outer medium (n_0) according to

$$n_0 \sin \alpha = (n_1^2 - n_2^2)^{1/2} \tag{6.2}$$

The left term of Eq. (6.2) is frequently referred to as the numerical aperture of a fiber. It is equal to the sine of α when the outer medium is air with its n_0 of 1.000. It numerical value ranges from 0.18 to 0.66. Since most multimode fibers rotate the light beam in a helical manner, they act as depolarizers. This can minimize errors attributed to polarization effects, but limits the capability of fibers for use in polarization experiments. Recently, however, the Hitachi Cable Co. has started offering single polarization monomode fibers which sustain polarization over distances as large as 100 m from 630 to 1300 nm.

A variety of fiber types is available, but fiber bundles have so far been used most often for sensing purposes. In a bifurcated fiber bundle (see Figure 4), light is focused into one bundle of lightguides which, at the sensing end, is statistically mixed with a second fiber bundle. Reflected light and/or fluorescence enters the strands of the second bundle and is guided to an optical separation and detection system. It is obvious from Figure 4 that there is a certain critical distance to be maintained in order to guarantee a sufficient overlap of the cones of incoming and outgoing fibers. Otherwise the outgoing fibers are unable to accept reflected light or fluorescence produced by the fibers of the incoming strand. Other fiber types are the single fibers (where the analytical information is carried in both directions within one fiber; see Figure 3) and the monomode fiber (which are extremely thin and allow light to protrude in a single mode only).

III. SENSOR CLASSIFICATION

Sensors may be classified either by their field of application or according to their working principle. The former classification includes four major types of FOCS, namely:

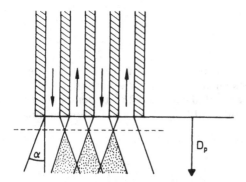

Figure 4   Schematic of a bifurcated fiber-optic bundle demonstrating the need for a critical distance (dashed line) to obtain overlap between the cones of incoming and outgoing strands. D_p is the depth of penetration of light into the medium to be sensed.

1. Physical sensors, capable of measuring physical parameters such as temperature or pressure
2. Chemical sensors, which measure the concentration of typical chemical species such as oxygen, carbon dioxide, ammonia, or pH
3. Enzyme-based biosensors, developed for species involved in biochemical reactions only, with typical examples being glucose, penicillin, or NADH sensors
4. Immunosensors, where the unique specificity of the antigen-antibody interactions is exploited and detected by optical means

The second classification is according to the working principle, and one may differentiate between intrinsic and extrinsic optical sensors. In the intrinsic sensor the information comes from the fiber, while in the extrinsic sensors the information comes from the sample or some sort of indicator chemistry at the end of the fiber. The extrinsic fiber sensors may be subdivided into

1. Bare-ended (first-generation) fibers
2. Indicator-phase sensors (second-generation fiber sensors)
3. Third-generation fiber sensors, i.e., those where a biocatalytic process is coupled to a first- or second-generation sensor.

The first sensors were of the intrinsic sensor type. Collection of information relied on the fact that alterations in a specific physical property of a medium being sensed would cause a predictable change in the light <u>transmission</u> characteristics of a fiber. In this case, the physical perturbation (such as a temperature change) interacts with the fiber to directly modulate the light intensity. Since in these devices the analytical information comes from the fiber itself, they are called intrinsic fiber sensors. Most of these lend themselves to measurement of physical parameters only. There is, however, a trend visible that intrinsic fibers will increasingly be used for chemical sensing too. A good example for such a device is a methane gas sensor in which the sensing element is a monomode fiber coated with platinum. The rate of the exothermic reaction of the hydrocarbon with oxygen is increased in the presence of the platinum catalyst. The resultant heat of reaction is transduced into a phase retardance in the light beam which is monitored [12].

Extrinsic sensors, in contrast, measure an optical parameter not associated with the fiber itself. Three possibilities exist: In the first one (first-generation sensors), the intrinsic optical properties of a medium are monitored. A representative example is the determination of the concentration of reduced nicotinamide adenine dinucleotide (NADH) using a bare-ended fiber-optic detection system [13]. NADH is produced in practically all cell media and displays a strong blue fluorescence when excited by UV light, for instance in an arrangement as shown in Figure 3. Since the concentration of NADH in a culture strongly depends on the oxygen partial pressure of the medium, the intrinsic fluorescence of NADH can be utilized as an indirect parameter for oxygen saturation.

However, only a limited number of chemical species (namely those having a color or fluorescence by themselves) can be detected by first-generation sensors. Moreover, these methods are of limited selectivity and selectivity in case of complex mixtures and mostly useful for making yes/no decisions only.

By combining the well-established indicator chemistry with fiber optics and immobilization techniques, a quite new technology has emerged in the past years that allows the construction of more selective and sensitive optical sensors. They are capable of measuring analytes via chemical or physical processes occurring at some sort of indicator chemistry at the fiber end, do no longer rely on the spectral properties of the sample and lend themselves to optical measurement of noncolored and nonfluorescent analytes such as oxygen, pH, or ammonia.

However, for some of the most important analytes no suitable indicators are known. In such a case, a biocatalytic system can be coupled to the indicator phase sensor by analogy to bioelectrodes. A typical example is provided by the glucose detection principle: Glucose can be oxidized by the enzyme glucose oxidase to give gluconic acid (H^+) and hydrogen peroxide under consumption of oxygen. Thus, by either monitoring the increase in pH [14], the production of H_2O_2, or the consumption of oxygen [15], a kinetic parameter for the glucose concentration is obtained.

Alternatively, the intrinsic fluorescence of a biocatalytic system may be monitored with a bare-ended fiber optic. Thus, NAD^+ is nonfluorescent, while its product of hydrogenation (NADH) is. As a result, the increase in fluorescence of a system consisting of, say, lactate dehydrogenase and lactate can be a parameter for the actual lactate concentration and therefore utilized in a biosensor [16]. Similarly, the intrinsic green fluorescence of glucose oxidase can be used to monitor glucose [17].

IV. SPECTROSCOPIC TECHNIQUES

The following optical parameters have been used so far in spectroscopic sensors: absorbance, reflectance, interferometry light scattering, fluorescence intensity, fluorescence lifetime, chemiluminescence, phosphorescence, thermoluminescence, and refraction index. Ellipsometry and polarimetry are two other useful methods, which, however, still await their application to fiber sensing.

A. Absorbance

Both the absorbance of the analyte and an immobilized indicator can be used. The relation between sample absorbance A and analyte concentration c is given by the Lambert-Beer law

$$A = \log \frac{I_o}{I} = \varepsilon cl \tag{6.3}$$

where A is the absorbance, c the concentration of the absorbing species, l the path length, ε the molar absorptivity, and I_0 and I are the intensities of incident and transmitted light, respectively. Absorbance-based measurements have been applied in optical sensing much less often than reflectance or fluorescence measurements, because most real samples are too strongly colored to allow precise absorbance measurements, and because absorbance methods cannot be adapted to turbid solutions.

B. Reflectance

This technique is rather popular in optical sensing. Both the reflectivity of a colored sample or that of an indicator layer may be measured. A typical example for the former method is the measure-

ment of blood reflectivity as a function of oxygen saturation:
Since oxygen causes a change in the absorption spectrum when
bound to hemoglobin, the measurement of the reflectivity of hemoglobin and oxyhemoglobin is a parameter for blood oxygenation.
The reflectance of immobilized phenol red as a function of pH, on
the other hand, can be exploited to monitor the pH of a sample
solution via optical fibers [18].

Reflection takes place when light infringes on a boundary surface, and two types of reflection are possible: The first is the
mirror-type or specular reflection, which occurs at the interface of
a medium with no transmission through it. The second (and more
realistic type in case of FOCS) is diffuse reflection, where the light
penetrates the medium and subsequently reappears at the surface
after partial absorption and multiple scattering within the medium.

Several models for diffuse reflectance have been proposed, all
of which consider that incident light is scattered by particles within
the medium. The most widely used is the Kubelka-Munk theory,
and an equation has been established that relates reflectance R with
the absorption coefficient K and the scattering coefficient S by

$$f(R) = \frac{(1 - R)^2}{2R} = \frac{K}{S} \qquad (6.4)$$

where $f(R)$ is the Kubelka-Munk function. The absorption coefficient K can be expressed in terms of the molar absorptivity ε and
the concentration c of the absorbing species, as $K = \varepsilon c$. Equation
(6.4) then becomes

$$f(R) = \frac{\varepsilon C}{S} = Kc \qquad (6.5)$$

where K is a constant since S is assumed to be independent of concentration. Equation (6.5) holds for a range of concentration for
solid solutions in which the absorber is incorporated with the scattering particles and for systems in which the absorber is adsorbed
on the surface of a scattering particle.

Ellipsometry is a polarimetric technique that measures changes
in the state of polarization of light upon reflection from a surface.
For a clean surface the optical constants of the surface and the
reflection coefficients of the system may be calculated from these
changes. Thin films on the surface cause additional changes from
which the thickness and refractive index of the film may be determined. The potential field of application for ellipsometry is in immunosensors.

C. Fluorescence

Among the variety of luminescence techniques available, fluorescence is certainly the most important one for optical sensing and shall be treated in this section in more detail. Fluorescence is the emission of light (or the emitted light) from an absorbing species that has been excited by light of different wavelength. The relation between fluorescence intensity and analyte concentration is given, in a first approximation, by

$$I_f = I_o k \phi_f \varepsilon l c \qquad (6.6)$$

where I_f is the intensity of fluorescent light, I_0 the intensity of incident light, k a constant related to the instrumental arrangement, ϕ_f the quantum yield, ε the molar absorptivity, l the path length in the sample, and c the concentration. The relation is linear up to absorbances around 0.5 and is negatively curved at higher absorbances.

Another useful equation in luminescence spectrometry is the Stern-Volmer equation, which relates the emission intensity of a luminophore with the concentration of a quencher:

$$\frac{I_o}{I} = 1 + k_{sv}[Q] = \frac{\tau_o}{\tau} \qquad (6.7)$$

In fluorescence, I_0 and I are the fluorescence intensities of a fluorophore in the absence and presence, respectively, of a quencher present in concentration [Q]; k_{sv} is the Stern-Volmer constant that is specific for each fluorophore-quencher combination and also depends on temperature and solvent. This equation is essential for all sensors based on dynamic quenching of fluorescence, for instance, in oxygen sensors.

It is obvious from Eq. (6.7) that not only light intensity, but also fluorescence lifetime (τ_0) is affected by a quencher in that it will be reduced to a smaller value τ. This is the basis for several new types of fluorosensors based on lifetime measurement rather than on light intensity measurement [19,20].

D. Scattering

Unlike reflectance measurement, scattering of light can also occur when there is no transition between ground and electronically excited states. Particles with sizes that are small compared to the wavelength of radiation (i.e., practically all atoms and molecules) give rise to Rayleigh scattering. If the particles are large, the phenomenon is called Mie scattering, the intensity of which can be related to the concentration of the scattering particles.

When the incident radiation can promote vibrational changes which can alter the polarization of a molecule, some of the energy of the electromagnetic wave will be lost. Therefore, the frequency of the light scattered by these particles will be different from that of the incident light and will have much weaker intensity. Such a scattering is known as Raman scattering and can be observed best when intense light sources such as lasers are used. Raman scattering is also observed with optical fibers, and the intensity of the fiber Raman band can serve as an internal reference.

V. CHEMICAL SENSORS

In view of the number of optical chemical sensors that has been published in the past years it is certainly impossible to present all of them. Rather, a few typical representatives shall be discussed in some detail in order to provide an idea on the possibilities of the technique.

A. NADH Sensors

Monitoring fluorescent NADH is an excellent example for the application of first-generation (bare-ended) fiber sensors in biotechnology. NADH, when excited at 345 to 350 nm, has a native fluorescence with a maximum at around 450 nm. Since the native fluorescence of NADH in cells appears to be a function of biomass concentration and, in particular of oxygen supply, measurement of NADH fluorescence intensity can provide useful information on these parameters. The principle of intracellular NADH measurement via fluorescence is shown in Figure 5.

NADH is present in all living cells of bacteria, fungi, plants, or animals. Unfortunately, the specific amount of NADH per cell varies within a vast range, but the following microorganisms have successfully have been studied with a fiber-optic flourosensor [13, 21,22]: baker's yeast, Bacillus subtilis, Candida tropicalis, Escherichia coli, Sporotrichum thermophile, Penicillium chrysogenum, Streptomyces sp., and Zymomonas mobilis. Many others have been investigated by conventional NADH fluorometry. The fluorescence data obtained with whole cells are said to be primarily relative data. They can be related to different metabolic states, to fluorescence data of a standardized process, or to biomass concentration, but a quantitation in terms of moles NADH per cell is difficult to achieve. A comparison of the NADH content of a batch cultiva-

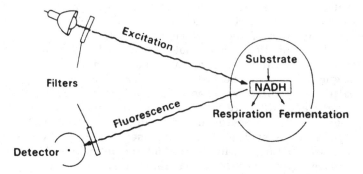

Figure 5 Principle of intracellular NADH measurement via fluorescence.

tion of baker's yeast as determined after cell rupture, with the NADH fluorescence signal obtained with a fiber inserted into the bioreactor is shown in Figure 6.

When, in contrast, the biomass is kept constant, the NADH fluorescence becomes an indicator for the metabolic state of the

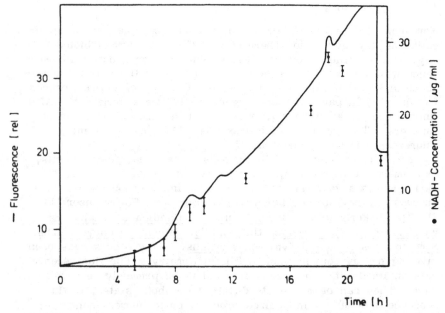

Figure 6 Relation between cell fluorescence at 350–450 nm, as observed with a fiber optic (full line), and the NADH concentration determined after cell rupture (data points). (From Ref. 13.)

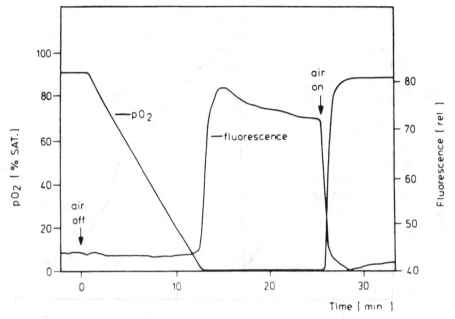

Figure 7   NADH-dependent fluorescence and oxygen partial pressure of a continuous culture of <u>Candida</u> <u>tropicalis</u> during aerobic-anaerobic transitions.   (Data from Ref. 13.)

cells.   This is also shown in Figure 6, where, during the initial phase, fluorescence corresponds closely to the NADH concentration, but at the end of the cultivation the signal drops quickly, indicating the depletion of the substrate and therefore the end of the process.

NADH serves as an electron donor to enzymes of the respiratory chain, and oxygen will therefore have an influence on its concentration.   Generally it is observed that when a cultivation is limited in oxygen, the NADH concentration is at least temporarily increased, before metabolism changes from aerobic to anaerobic.   The metabolic change can also be detected by measuring the NADH-dependent fluorescence.   An example is given in Figure 7, where the NADH fluorescence is seen to stay constant until oxygen pressure reaches a limiting level.   At this critical level, fluorescence rises fast, indicating that the oxygen supply is now too small to maintain aerobic metabolism.

Finally, monitoring of NADH fluorescence can also serve to study substrate-dependent phenomena in bioreactors, for instance

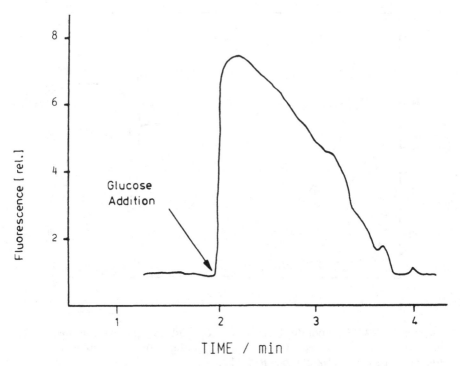

Figure 8 Fluorescence signal of NADH after addition of glucose to substrate-limited baker's yeast cells. (Data from Ref. 23.)

the kinetics of substrate uptake [23,24], without cell rupture. Figure 8 illustrates the effect of substrate addition to substrate-limited cells in a cultivation of baker's yeast.

B. Gas and pH Sensors

In contrast to NADH, most gases and, of course, the proton, are colorless and therefore cannot be assayed with first-generation fiber sensors. Therefore, advantage is taken of the well-established indicator chemistry for a variety of species. Oxygen, for instance, is known as a notorious quencher of the fluorescence of aromatic compounds. Carbon dioxide, on the other hand, can be assayed by analogy to the CO_2-sensitive electrodes by using a buffer system entrapped in a polymer: The pH of the alkaline buffer undergoes reversible changes according to $OH^- + CO_2 \rightarrow HCO_3^-$, and the resulting decrease in pH is indicated by a pH-sensitive dye whose color or fluorescence is seen by the fiber. A schematic of such a sensor is shown in Figure 9.

In a series of papers, Lübbers and Opitz [2,25,26] have demonstrated the feasibility of optical sensors (called "optrodes" [25] or "optodes" [26] by lingual analogy to the electrodes) for carbon dioxide and oxygen. Sensors for these species have numerous applications and, in addition, can be exploited as transducers for biochemical reactions involving one of the above-mentioned species. In recent years, a number of papers on sensors for oxygen [20,27–31], ammonia [32] and carbon dioxide [33,34] have been published.

Kroneis and Marsoner [31] have developed a sterilizable fiberoptic oxygen probe specifically for use in bioreactors and have compared its performance with an oxygen electrode. The most characteristic features of the optical sensor are good long-term stability, lack of interference by carbon dioxide, and fairly quick response time.

We have recently reported [20] on a quite new type of oxygen sensor with the fluorescence decay time as the information carrier. A blue LED is used as a light source, and a long-lived oxygen indicator is periodically excited with its light. The phase shift between fluorescence and phase-modulated excitation can be related to oxygen partial pressure and gives a calibration graph that is distinctly more linear than the corresponding graphs obtained with intensity measurement based sensors. The advantage of such a system is that it has an internal reference system (namely, the excitation sine) so that it displays a very good long-time stability.

The high information density flux that can be maintained within a fiber has been utilized to develop a single sensor for two analytes, namely, oxygen and carbon dioxide [34]. A schematic of such a sensor is shown in Figure 10. The CO_2-sensitive layer (CSL) consists of a buffer solution entrapped in a gas-permeable polymer and contains a pH-sensitive indicator. The oxygen-sensitive layer (OSL) is atop of the CSL and contains a ruthenium dye in kieselgel beads entrapped in another polymer. The upper layer (OI) is the optical isolation that prevents ambient light and sample fluorescence to enter the fiber. Both indicators have the same wavelength of excitation, but quite different emissions. Thus, the green fluorescence of the CSL is a signal for CO_2, and the red fluorescence of the OSL a signal for oxygen (Figures 11 and 12).

Sensors for pH have been developed by various groups [18,35–38]. In each case, a pH-sensitive material is immobilized at the end of a fiber. In contact with the sample, the indicator suffers a change in color or fluorescence that can be related to pH. Practically all sensors can measure over the physiological pH range only.

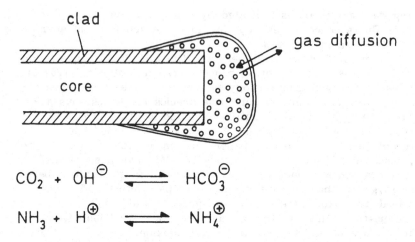

$$CO_2 + OH^{\ominus} \rightleftharpoons HCO_3^{\ominus}$$

$$NH_3 + H^{\oplus} \rightleftharpoons NH_4^{\oplus}$$

Figure 9 Schematic of a fiber sensor tip useful for sensing acidic or basic gases. The gas diffuses through the polymer and causes a change in pH in a buffer solution entrapped as an emulsion in the polymer. The fiber "sees," within its numerical aperture, the change in the color of an indicator added to the aqueous emulsion.

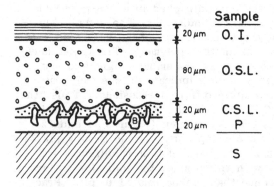

Figure 10 Cross section through a sensing layer at the end of a glass fiber and capable of measuring oxygen and carbon dioxide simultaneously in a gaseous or fluid sample.

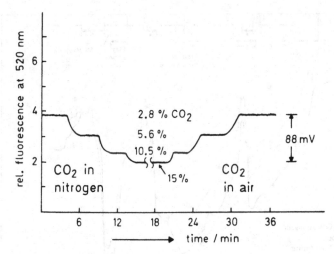

Figure 11 Response of the O_2/CO_2 optrode toward carbon dioxide in nitrogen and air by measuring the fluorescence at 520 nm.

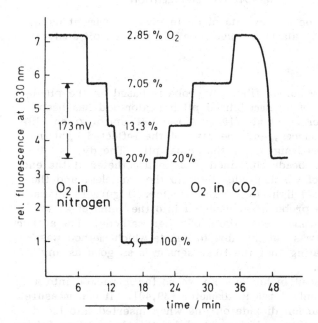

Figure 12 Response of the O_2/CO_2 sensor toward oxygen in nitrogen and CO_2 by measuring the fluorescence of the sensor at 630 nm.

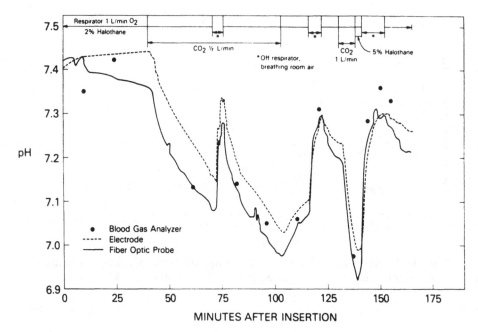

MINUTES AFTER INSERTION

Figure 13   Comparison of the data of an in vivo experiment using a
pH optrode with the data of an electrode and a blood pH analyzer.
(From Ref. 18.)

The first ever reported fiber pH probe is based on the pH-de-
pendent reflectivity of an immobilized pH indicator and has been
constructed by Peterson et al. [18].   A red LED and a green LED
are used as light sources, and the ratio of the reflected light intensity is a normalized signal for the actual pH.   The dye is adsorbed onto polymer beads contained in a small catheter at the end
of two fibers, one of which guides light to the particles, and the
other guides reflected light to a photodetector.   Figure 13 shows
the response of the probe when inserted into the veins of a ewe
and how the data compare with those of other methods.   The authors
have stated that it was not possible to decide which method was
most precise, indicating that the fiber sensor is as good as any
conventional device.

The most advanced technology known so far is packed into a
small triple sensor only 1 mm in diameter [39,40].   It can measure
pH, oxygen, and carbon dioxide on-line when inserted into blood
vessels during operations.   Figure 14 shows the tip of this triple
sensor, consisting of three 100-μm-thick fibers with three different

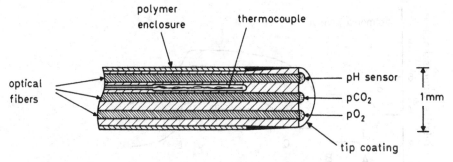

Figure 14   Cross section of a 1-mm fiber-optic catheter for continuous monitoring of oxygen, carbon dioxide, and pH. (From Ref.
39.)

analyte-sensitive layers at their end, along with a conventional
thermocouple for measuring temperature.   All three parameters are
determined by fluorimetry: oxygen via its quenching effect upon
an aromatic hydrocarbon, pH via the changes in fluorescence of a
covalently immobilized indicator, and carbon dioxide via optically
detected changes in the pH of a buffer solution entrapped in a
polymer.

A first version of this sensor ("Gas-Stat") that can monitor
blood gases and pH in an extracorporeal loop has been commercially
available since 1984.   The system consists of a microprocessor-based
instrument, bifurcated fiber-optic cables, and a disposable sensor
head with fluorescent spots sensitive to the respective analytes.

Ammonia may be sensed in a fashion similar to CO_2, in that an
added indicator "sees" the increase in pH when ammonia reacts with
a buffer according to $H^+ + NH_3 \rightarrow NH_4^+$, thereby raising the pH.   A
fiber tip emulsion similar to the one shown in Figure 9 was used in
the first fluorosensor for ammonia [32], with a pH 8 buffer entrapped as a fine emulsion in silicone rubber.   The indicator itself
acts as a buffer.   Figure 15 shows the relatively slow response of
the sensor.   In a reflectance-based ammonia sensor, spectral
changes of nitrophenol with pH have been used to monitor ammonia
[41].

C.   Electrolytes

The sensing schemes for ions include complexation of indicators by
metal ions, ion-pair extraction, and ion-carrier mediated sensing.
An interesting example of a sensor based on complexation of a chro-

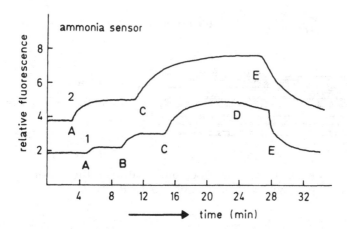

Figure 15 Response of an ammonia optrode. Curves: (1), 1-naphthol-sulfonate in ammonia buffer; (2) 0.3% naphtholsulfonate in water, adjusted to pH 8.2. Starting with water, the sensor was exposed to (A), 0.1-m$\underline{M}$; (B), 0.5-m$\underline{M}$; (C), 1-m$\underline{M}$ ammonia solutions. Washing was done with (D) pure water and (E) 0.1 $\underline{M}$ hydrochloric acid.

mogenic crown ether by potassium has been described by Alder et al. [42]. Zhujun et al., on the other hand, have developed a method based on competitive binding and ion-pair extraction in order to measure sodium ion [43]. The third detection principle is based on ion carriers and fluorescent dyes sensitive to a potential formed at a lipid-water interface [44]. The method is similar to the one applied in ion-selective electrodes, except that the potential is measured optically.

VI. BIOSENSORS

Two types of biosensors are known: In the first one, a fiber-optic chemical sensor (such as an oxygen optrode) is coupled to a biocatalytic system that changes the concentration of a species (such as oxygen). In the second, changes in the intrinsic fluorescence of a biosystem are monitored with a plain fiber. A few representative examples shall be given in order to demonstrate the scope of the method.

A. Glucose

Optosensors for glucose are based on the measurement of oxygen consumed as a result of the enzymatic oxidation of glucose by glu-

cose oxidase in the presence of catalase [15]. The experimental
arrangement is analogous to the one shown in Figure 16, in that an
enzymatic reaction occurs in a layer close to an oxygen-sensitive
layer whose fluorescence is monitored. The oxygen optrode (based
on fluorescence quenching) acts as a transducer for the enzymatic
reaction. Aside from glucose, lactate and cholesterol [45] may also
be assayed using the same principle (for a review, see Ref. 5).

Another type of glucose sensor exploits the decrease in pH as a
result of glucose oxidation to gluconic acid [14]. A promising approach for glucose [46] is based on competitive binding of glucose
and a competing ligand (FITC-labeled dextran) for concanavalin A
immobilized at the end of a fiber. The numerical aperture of the
fiber sees only unbound dextran, the amount of which increases
with the amount of glucose being present. This sensor principle is
of particular interest because it has a broad potential for applicat-
tion. It may be adapted to any analytical problem for which a specific competitive binding system can be devised.

B. Penicillin

Fuh and coworkers [47] as well as Kulp et al. [48] have developed
enzyme optrodes for the detection of penicillin. In a typical arrangement, a polymer membrane is covalently attached to the tip of
a glass optical fiber. The membrane contains the enzyme penicillinase and a pH-sensitive fluorescent dye. The enzyme catalyzes
the cleavage of the lactam ring of penicillin to produce penicilloic
acid and, consequently, a pH change in the microenvironment. Response times are about 1 min, and linear response over the 0.25 to
10 mM penicillin G concentration range is observed. Of course, the
response function depends on the pH of the sample solution and its
buffer capacity.

C. Other Biosensors

Ethanol may be assayed [49] using a fiber sensor whose sensitive
layer is schematically shown in Figure 16. Ethanol diffuses through
the polymer membrane into small particles in the membrane which
contains an aqueous solution of an oxygen-sensitive dye and the
enzymes alcohol oxidase and catalase. Upon oxidation of alcohol to
acetaldehyde, oxygen is consumed, which is indicated by the oxygen
sensor. Ethanol in the 0.1 to 1.5 mol/L concentration range has
been found to be detectable. The advantage of such as system lies
in the fact that enzyme and indicator are fully protected from the
analyte, so that interferences by pH and heavy metals can be neg-
lected. On the other hand, the response time of the sensor is pro-
longed to 2—5 min because of the diffusional processes involved.

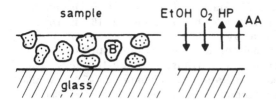

Figure 16 Construction of an ethanol biosensor with an oxygen op-trode as a transducer, and diffusion processes involved. Kieselgel beads containing catalase, alcohol oxidase, and oxygen-sensitive indicator and embedded in a 100-μm-thick layer of silicone rubber on the glass. On the right side, the chemical species involved in the biocatalytic system are given along with the direction of their diffusion. EtOH, ethanol, HP, hydrogen peroxide, AA, acetaldehyde.

Lactate has been determined using immobilized lactate oxidase as the biocatalyst and an oxygen optrode as the transducer which measures the decrease in oxygen tension [50]. Two oxygen-sensing layers were used in order to account for the varying background oxygen level that supplies the biocatalytic system. In a different design for a lactate biosensor the intrinsic fluorescence of a biocatalytic system at the distal end of a fiber has been utilized [50]: When lactate dehydrogenase oxidizes lactate to pyruvate, NADH will be produced, which can be monitored fluorimetrically. A similar approach enabled the determination of ethanol via alcohol dehydrogenase. Finally, the immobilization of deaminating enzymes at the fiber tip together with an ammonia sensor allows the monitoring of biocatalytic reaction during which ammonia is produced or consumed [51].

D. Enzyme Activity Probes

A fiber-optic sensor for the direct kinetic determination of enzyme activities [52] is based on the immobilization of chromogenic or fluorogenic substrates at the end of a fiber. When hydrolyzed by an enzyme, the substrate is converted into a colored and highly fluorescent product. The increase in fluorescence with time can be correlated to the activity of the enzyme (an esterase). It should be kept in mind that this type of probe is not a sensor, since it acts irreversibly. To differentiate it from a true sensor, it is suggested to use the term "probe" exclusively for this type of devices, and to confine the word "sensor" to devices that act <u>continuously</u> <u>and</u> <u>reversibly</u>.

The method may be extended to other hydrolases, such as phophatases, sulfatases, or glucosidases. The synthetic substrates

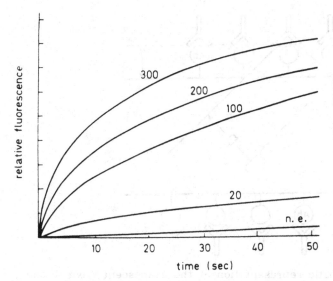

Figure 17 Response of a fiber probe for the enzyme carboxyles-terase. The figures refer to micrograms of enzyme per milliliter solution, into which the probe was immersed, and n.e. is the sig-nal due to nonenzymatic hydrolysis.

do not present a health risk when applied in vivo since it is the nontoxic acid or sugar component that is released, whereas the dye remains immobilized. The probe can be used several times: for instance, first in a calibration step, and then for sample sensing. The sensitive layer at the fiber end is preferentially a disposable thin layer of immobilized substrate. Figure 17 shows how the signal develops with time together with the rate of nonenzymatic hydroly-sis for comparison.

VII. IMMUNOSENSORS

An immunosensor is a device capable of quantitatively detecting the concentration of an antibody, hapten, or antigen. Aside from el-lipsometry, internal reflection techniques (IRTs) including absorb-ance, fluorescence, and light scattering techniques have been ap-plied most frequently either on optical planar waveguides or with fibers [52]. All kinds of IRT are based on the fact that light, up-on total reflection at an interface, penetrates ("evanesces"), a characteristic distance (typically 100 μm) into the optically rarer me-dium before it returns into the optically denser one. The evanescing

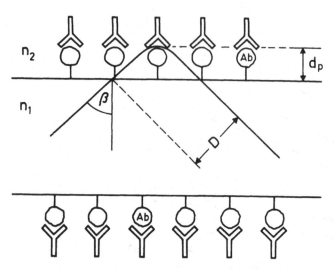

Figure 18 Schematic representation of the evanescent wave at the interface of phases of different refractive index. Light penetrates into the phase with n_2 to a depth d_p and can hit an absorbing molecule there, for instance a labeled antibody (Ab). D is the phase shift between incident and returning beam (the Goos-Hänchen shift).

wave may be absorbed in the second phase (which, for instance, is a dye-labeled protein; see Figure 18) or may produce fluorescence in the outer medium. Thus, IRTs are obviously well suited for probing optical properties of thin layers on a transparent substrate or a waveguide.

Most work on total internal reflection was performed with fluorescence detection. It is very useful to read Tomas Hirschfeld's work [53,54]. At a very early stage, Hirschfeld pointed out the advantages of the small penetration depth (d_p) of the evanescent field for turbid or highly absorbing solutions, the efficient excitation of thin-film layers by multiple reflections, and how variation of d_p can be achieved by simply varying the incident angle and wavelength.

Andrade et al. [55] utilized the intrinsic fluorescence of tryptophan in proteins to generate a signal change following adsorption of bovine serum albumin or bovine-gamma-globulin to a quartz surface. Unfortunately, the intrinsic fluorescence of proteins is at rather short wavelengths (excitation at ca. 280–295 nm, fluorescence ca. 320–350 nm), so expensive quartz waveguides have to be employed.

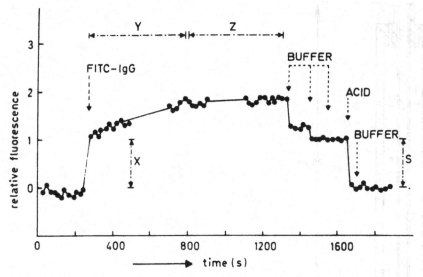

Figure 19 Specific binding of labeled IgG by antiserum at a glass-liquid interface. The binding process is monitored by the increase in fluorescence which is excited by the evanescent wave of the beam propagating in the slide.

It has therefore been tried to label the protein with a long-wave-absorbing tag [56]. In a typical sandwich immunoassay experiment, immobilized sheep antiserum was incubated with antigen to form one-half of the sandwich. After washing out excess IgG, the solid-phase complex was reacted with FITC-labeled antiserum, and the reaction was monitored for 15 min. Figure 19 shows the detection of the binding process. Following the injection of labeled anti-IgG, a rapid increase in fluorescence intensity is observed (X). This is due mainly to free molecules fluorescing within the penetration depth of the evanescent wave. Second, a slower increase in binding over the next 500 s is evident, reaching a plateau after approximately 600 s. The curve obtained when no antigen was present shows fluorescence because of the presence of free fluorescent molecules, but it lacks the binding event.

Hirschfeld [54] as well as Tromberg et al. [57] have devised immunoassays using optical fibers. A schematic of a typical device is shown in Figure 20. The fiber (A) is covered with a fluorescent antigen-antibody couple and immersed into a sample volume containing the unknown concentration of the same antigen. By equilibration part of the fluorescent antibody will be replaced by the non-

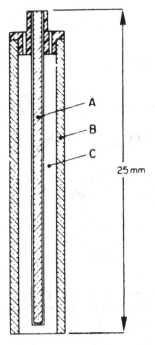

Figure 20 A disposable fiber-optic immunosensor. The lightguide (A) is immersed into a capillary tube (B) which defines the sample volume (C).

fluorescent sample antigen, resulting in a decrease of the fluorescence at the surface of the fiber. Fluorescence intensity is measured by the evanescent wave technique which can induce fluorescence only at sites close to the fiber surface, but not in bulk solution.

Various other optical methods have been exploited for immunoassay on waveguides and are summarized in Ref. 58. It is probably informative to mention the use of light scattering as another sensing technique that can help to avoid labeling of proteins. Thus, the antibody to human IgG was adsorbed onto the surface of a quartz slide and 410-nm light guided within the slide. As the antibody binds to the antigen, there is an increase in scattered light detectable perpendicular to the surface (Figure 21). Kinetic monitoring was possible, and the system was sufficiently sensitive to differentiate 20 µg/mL IgG from the background [59].

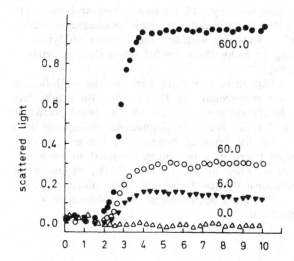

Figure 21 Generation of light scattering at the surface of an optical waveguide bearing immobilized antibody. IgG was added after 2 min in concentrations ranging from 0.0 to 60.0 µg/mL, and scattering monitored at 440 nm. (Data from Ref. 59.)

VIII. OUTLOOK

The design of new fiber sensors is a quite active area of research, and numerous promising ideas are likely to be pursued in the future. Aside from conventional intensity measurements, other parameters will be used: for instance, fluorescence decay time [19, 20] rather than intensity, fluorescence polarization, or energy transfer (ET). The latter will provide quite new possibilities and a greater flexibility in the choice of dyes, particularly for pH [60] or oxygen [61]. Another field of application of ET-based sensors is the investigation of immunological binding reactions, since the efficiency of dipole-dipole ET decreases with the sixth power of distance. Energy-transfer-based sensors can be divised for both ground-state and excited-state processes, and both trivial (reabsorption) and dipole-dipole mechanisms of ET offer certain advantages.

Aside from new methods in spectroscopy and immobilization techniques, progress will be achieved by improvement of optical components. Inexpensive lasers are available in the near-IR, and LEDs can be used instead of using light sources requiring more power, in many cases. In addition to conventional fiber-optic techniques (with measurements performed at the distal end of the waveguide) evanescent wave spectrometry will be used more and more in order to improve signal intensities and to monitor processes occurring at interfaces. Finally, existing sensors for various analytes need considerable finishing to make them useful when dealing with real samples in biotechnology.

It is hoped that this chapter can mediate some of the enthusiasm that is currently shared by researchers in this field. We think that fiber-optic sensors can significantly contribute to the development of new sensing techniques and that, for entrepreneurs, the opportunity exists to participate in a rapidly growing segment of the market for chemical analyses, although there may be tough competition from other sensor instrumentation, such as electrodes, FETs, piezoelectric devices, Fourier transform IR spectroscopy, and calorimetry. Most probably, an interdisciplinary approach to the solution of existing hurdles will be essential.

REFERENCES

1. Turner, A. P. F., Karube, I., and Wilson, G. S. (eds.), Biosensors. Fundamentals and Applications, Oxford University Press, New York (1987).
2. Lübbers, D. W. and Opitz, N., Sensors & Actuators, 4, 641 (1984).
3. Peterson, J. I. and Vurek, G. G., Science, 224, 123 (1984).
4. Hirschfeld, T., Callis, J. B., Kowalski, R. B., Science, 226, 312 (1984).
5. Wolfbeis, O. S. (ed.), Fiber Optic Chemical Sensors and Biosensors, CRC Press, Boca Raton (FL) (1990).
6. Seitz, W. R., CRC Crit. Rev. Anal. Chem. 19, 135 (1988).
7. Wolfbeis, O. S., Pure & Appl. Chem., 59, 662 (1987).
8. Alder, J. F., Fresenius Z. Anal. Chem., 324, 372 (1986).
9. Schmidt, R. D. (ed.), Proceedings of the Biosensors International Workshop, GBF Monograph, Vol. 10. VCH Verlag, Weinheim (1987).
10. Edmonds, T. E. (ed.), Chemical Sensors, Blackie, London (1988).
11. Wolfbeis, O. S., in Molecular Luminescence Spectrometry. Methods and Applications (S. G. Schulman, ed.), Wiley, New York, (1988), Vol. 2, Chap. 3.
12. Farahi, F., Akhavan-Leilabady, P., Jones, J. D. C., and Jackson, D. A., J. Phys. E: Sci. Instrum., 20, 435 (1987).
13. Beyeler, W., Eisendle, A., and Fiechter, A., Eur. J. Microbiol. Biotechnol., 13, 10 (1981).

14. Trettnak, W., Leiner, M. J. P., and Wolfbeis, O. S., Biosensors 4, 15 (1989).
15. Trettnak, W. and Wolfbeis, O. S., Analyst 113, 1519 (1988).
16. Wangsa, J. and Arnold, M., Anal. Chem. 60, 1080 (1988).
17. Trettnak, W. and Wolfbeis, O. S., Anal. Chim. Acta 221, 195 (1989).
18. Peterson, J. I., Goldstein, R. V., and Buckhold, D. K., Anal. Chem., 52, 864 (1980).
19. Vickers, G. H., Miller, R. M., and Hieftje, G. M., Anal. Chim. Acta, 192, 145 (1987).
20. Lippitsch, M. E., Pusterhofer, J., Leiner, M. J. P., and Wolfbeis, O. S., Anal. Chim. Acta 205, 1 (1988).
21. Meyer, H. P., Beyeler, W., and Fiechter, A., J. Biotechnol., 1, 341 (1984).
22. Scheper, T., Lorenz, T., Schmidt, W., and Schügerl, K., J. Biotechnol. 3, 231 (1986).
23. Eisendle, A., Ristroph, D. L., and Humphrey, A., Eur. J. Appl. Biotechnol., 6, 335 (1979).
24. Beyeler, W. and Meyer, C., Biotech. Bioeng., 26, 916 (1984).
25. Lübbers, D. W. and Opitz, N., Pflüger's Arch., 359, R145 (1975).
26. Lübbers, D. W. and Opitz, N., Z. Naturforsch., 30C, 532 (1975).
27. Wolfbeis, O. S., Offenbacher, H., Kroneis, H., and Marsoner, H., Mikrochim. Acta, I, 153 (1984).
28. Peterson, J. I., Fitzgerald, R. V., and Buckhold, D. K., Anal. Chem., 54, 62 (1984).
29. Wolfbeis, O. S., Posch, H. E., and Kroneis, H. K., Anal. Chem., 57, 2556 (1985).
30. Wolfbeis, O. S., Leiner, M. J. P. H., and Posch, H. E., Mikrochim. Acta III, 359 (1986).
31. Kroneis, H. W. and Marsoner, H. J., Sensors & Actuators, 4, 587 (1983).
32. Wolfbeis, O. S. and Posch, H. E., Anal. Chim. Acta, 185, 321 (1986).
33. Zhujun, Z. and Seitz, W. R., Anal. Chim. Acta, 160, 305 (1984).
34. Wolfbeis, O. S., Weis, L., Leiner, M. J. P., and Ziegler, W., Anal. Chem. 60, 2028 (1988).
35. Zhujun, Z. and Seitz, W. R., Anal. Chim. Acta, 160, 47 (1984).
36. Offenbacher, H., Wolfbeis, O. S., and Fürlinger, E., Sensors & Actuators, 9, 73 (1986).
37. Fuh, M. R. S., Burgess, L. W., Hirschfeld, T., and Christian, G. D., Analyst, 112, 1159 (1987).

38. Grattan, K. T. V., Mouaziz, Z., and Palmer, A. W., Biosensors.

39. Gehrich, J. L., Lübbers, D. W., Opitz, N., Hansmann, D. R., Miller, W. W., Tusa, J. K., and Yafuso, M., IEEE Trans. Biomed. Eng., 33, 117 (1986).

40. Miller, W. W., Yafuso, M., Yan, C. F., Hui, H. K., and Arick, S., Clin. Chem., 33, 1538 (1987).

41. Arnold, M. A. and Ostler, T. J., Anal. Chem., 58, 1137 (1986).

42. Alder, J. F., Ashworth, D. C., Narayanaswamy, R., Moss, R. E., and Sutherland, I. O., Analyst, 112, 1191 (1987).

43. Zhujun, Z., Mullin, J. L., and Seitz, W. R., Anal. Chim. Acta, 184, 251 (1986).

44. Wolfbeis, O. S. and Schaffar, B. P. H., Anal. Chim. Acta, 198, 1 (1987).

45. Trettnak, W. and Wolfbeis, O. S., Anal. Lett. 22, 2191 (1989).

46. Schultz, J. S., "Optical Sensor for Plasma Constituents," U.S. Patent 4,344,438 (1982).

47. Fuh, M. S., Burgess, L. W., and Christian, G. D., Anal. Chem. 60, 433 (1988).

48. Kulp, T. G., Camins, I., Angel, S. M., Munkholm, C., and Walt, D. R., Anal. Chem. 59, 2849 (1987).

49. Wolfbeis, O. S. and Posch, H. E., Z. Anal. Chem. 332, 255 (1988).

50. Trettnak, W. and Wolfbeis, O. S., Anal. Lett., 22, 2191 (1989).

51. Arnold, M. A., in Proceedings of the Biosensors International Workshop, GBF Monograph, Vol. 10, VCH Verlag, Weinheim (1987), pp. 223–227.

52. Wolfbeis, O. S., Anal. Chem., 58, 2874 (1986).

53. Hirschfeld, T., Can. J. Spectr., 10, 128 (1965).

54. Hirschfeld, T., "Fluorescent Immunoassay Employing Optical Fiber in Capillary Tube," U.S. Patent 4,447,546 (1984).

55. Andrade, J. D., VanWagenem, R. A., Gregoris, D. E., Newby, K., and Lin, J. N., IEEE Trans. Electr. Dev., 32, 1175 (1985).

56. Sutherland, R. M., Dähne, C., Place, J. F., and Ringrose, A. S., Clin. Chem., 3, 1533 (1984).

57. Tromberg, B. C., Sepaniak, M. J., Vo-Dinh, T., and Griffin, G. D., Anal. Chem., 59, 1226 (1987).

58. Place, J. F., Sutherland, R. M., Dähne, C., Biosensors, 1, 321 (1985).

59. Sutherland, R. M., Dähne, C., and Place, J. F., Anal. Lett., 17, 43 (1984).

60. Jordan, D. M., Walt, D. R., and Milanovich, F. P., <u>Anal. Chem.</u>, <u>59</u>, 437 (1987).
61. Sharma, A. and Wolfbeis, O. S., <u>Appl. Spectrosc.</u> <u>42</u>, 1009 (1988).

7

Applications of Electrochemical Sensors

HANS W. BUEHLER and RENÉ BUCHER   *Ingold Messtechnik AG, Urdorf, Switzerland*

I. INTRODUCTION

This chapter details the use of conventional electrochemical sensors for the measurement of pH, dissolved oxygen, redox potential, and dissolved carbon dioxide. It mainly covers in situ sensors, which are mounted directly in the bioreactor. First, the principal requirements for in situ sensors must be discussed.

Looking back over the past 30 years, one sees the development of only a few in situ chemical sensors. The main reason for this is the need for sterilizability of such a sensor. In other words, the sensor has to withstand steam of up to 130°C and up to 2 bar pressure. These harsh conditions represented a big hurdle for the development of such sensors. Experience has shown that the problems in sensor development increased exponentially with an increasing upper temperature limit. Happily the present upper limit of 130°C for sterilizable sensors will not change in the near future.

Besides temperature and pressure, bioprocesses pose other problems for the sensor. A fundamental requirement of sensors, mounted in the bioreactor, is the long-term stability of the zero point and slope. A recalibration during the cultivation is in most cases complicated and time-consuming or even impossible. Commonly the in situ sensor is checked by taking a sample. However, the sampling and subsequent analysis in the laboratory necessitates very careful and reproducible handling.

Maintaining sterility is another prerequisite for a successful cultivation. Hence the mounting and design of the sensor have to be optimized. Incorrect positioning of an O-ring alone may lead to an infection of the culture. Some sensors are completely closed (O_2, CO_2), others must have an electrolytic contact with the broth for functioning (pH, redox) which may be critical.

Another characteristic of bioprocesses is the dynamic nature of microorganisms changing themselves and their environment with time.

Yet microorganisms do not violate the fundamental laws of thermo-
dynamics. However, there is never a thermodynamic equilibrium between the liquid and gas phases. Hence, in aerated and mixed
bioreactors, the O_2 partial pressures or activities in the gas and
liquid phases are not identical. The same holds for the CO_2 partial
pressure. The O_2 and CO_2 sensors, however, are in contact with
solution and gas bubbles. The sensor should be designed to analyze only the liquid phase. The gas phase is measured by off-gas
analyzers. This problem does not exist for pH and redox electrodes.

The flow dependence of a sensor signal is also an important
topic. In order to understand it, the following principal classes of
sensors have to be differentiated:

Type A: Potentiometric sensors which do not consume anything
          from the broth (pH, redox, CO_2)
Type B: Amperometric sensors which continuously consume some-
          thing from the broth (O_2).
          On-line sensors using continuous sampling and analyzing also
          belong to this class.

Flow dependence exists only with type B sensors. Correct results
demand a certain minimal flow velocity, which is never a problem in
well-aerated systems. In the cultivation of animal or mammalian
cells, however, the flow dependence of sensors is a real drawback.
In addition, the flow dependence increases with increasing viscosity
of the medium.

Another characteristic of bioprocesses is the complexity of the
fermentation medium. Hence, highly selective sensors need to be
developed.

The field of biotechnology has made substantial progress within
the last decade. The advances made by molecular biologists and
genetic engineers have led to a new generation of bioprocesses.
Cell yields and productivity have been greatly improved. However,
these processes are more delicate and need better control compared
to traditional cultivations. New processes using animal or human
cell lines for the production of high-value products such as monoclonal antibodies have become more and more important. Corresponding bioreactors are small and hence ask for small and accurate
sensors. A further aspect is the growing importance of computer-
controlled bioprocesses, where again stable and accurate sensors are
highly desired. It is stated that one of the bottlenecks that limits
the application of computers is the lack of suitable sensors. It follows from the above statements that more accurate and smaller sensors are required.

A last but important parameter is the maintenance and handling
of a sensor. Experience has shown that the operator is usually un-

Table 1 Specifications for in Situ Sensors

Characteristics of bioprocesses	Requirements
Sterilization	Good thermal stability and compressive strength, no temperature hysteresis
Sterility	Optimal sensor design, fully sterilizable
Long-term process	Long-term stability of calibration values
Varying flow velocity	No flow dependence
Complex media	No interference
Multiphase system	No interference by air bubbles (O_2 and CO_2) or by microbes
Cell cultures	Small and accurate sensors
Maintenance	Easy handling

able to realize the sensor specifications if the handling is tricky. Table 1 summarizes the foregoing statements.

It is the aim of this chapter to analyze critically the present sensors according to Table 1. The theory of the sensors is briefly described in order to understand the current problems. The full theory of the individual sensors is complicated and beyond the scope of this book. The corresponding bibliographical sources are mentioned.

II. pH MEASUREMENT

A. Theoretical Aspects

The pH value is commonly measured with a combined glass-reference electrode. The steam sterilizable pH sensor was developed by Ingold [1] and became a standard device in the fermentation industry. Later Leeds & Northrup (L&N) introduced a sealed sterilizable pH sensor based on a gelled, instead of a liquid, electrolyte. Recently other manufacturers announced pH electrodes which appear to have similar specifications comparable to either the Ingold or L&N electrodes. Due to extensive experience, the following descriptions mainly deal with the Ingold electrode. Figure 1 shows the principal parameters of a combined pH sensor with a sterilizable electrode head.

Figure 1 Sterilizable Ingold electrode: 1, pH glass membrane; 2, internal buffer; 3, lead-off electrode of the glass electrode; 4, diaphragm; 5, high-temperature Ag/AgCl electrode; 6, reference electrode.

The pH-sensitive glass reacts with water, which results in a gel layer. Within this layer there arises the potential, which is dependent on the activity of hydrogen ions in the solution. The potential on the inside of the glass membrane is kept constant by a buffered internal solution.

The potential of the reference electrode should be constant for an extended period of time. Its stability is ensured by a constant chloride concentration in the reference electrolyte. Any evaporation or contamination must be avoided. For this reason, an overpressure in the reference electrode is favorable. The overpressure guarantees a small flow of electrolyte through the diaphragm and thus prevents test solution from diffusing into the reference electrode. Very often a slightly viscous KCl solution is used in order to have a reasonably small outflow of electrolyte. The L&N electrode and others have a gelled electrolyte and no refilling hole. An overpressure cannot be applied and therefore the contamination of the electrolyte is possible. Electrodes with a bridge electrolye have two electrolyte chambers. The outer chamber contains the bridge electrolyte (usually a KCl solution) which protects the Ag/AgCl electrode in the inner chamber from contamination. This protection, however, is not perfect due to the diffusion of substances from the bridge electrolyte through the inner diaphragm.

The diaphragm serves to establish electrolytic contact between the test solution and the reference electrolyte. A small potential is always created at the diaphragm. It must be small and constant in order to accurately measure pH values.

The potential U of a combined pH electrode obeys Eq. (7.1):

$$U = E_g - E_r - E_d - s\, pH \qquad (7.1)$$

where

E_g = zero potential of the glass electrode

E_r = potential of the reference electrode

E_d = diaphragm potential

s = slope (temperature dependent, i.e., 60.6 mV/30°C)

The most important prerequisite for a sterilizable pH electrode is the reversibility of all temperature related physical/chemical effects. Such ideal electrodes do not exist due to the slow aging of the gel layer of the glass membrane. The calibration values therefore change slightly because of steam sterilizations. However, these changes are negligible with modern high-quality glass electrodes. The potential shift of the reference electrode depends on the type

Table 2   Characterization of Reference Electrodes

Liquid electrolyte pressurized	Gelled electrolyte nonpressurized
Fast response	Slow response when old
Possibility of using different reference or intermediate electrolytes	Only one electrolyte
High accuracy	Accuracy not necessarily high
No contamination of electrolyte	Contamination by measuring solution
Long lifetime	Medium lifetime

Maintenance
    pressurizing
    refilling
	No maintenance

of electrode. Having a liquid electrolyte and overpressure, the shift is negligible. The potential of gel-type electrodes slowly shifts due to KCl diffusion. Table 2 summarizes the major differences between the two main types of reference electrodes. A more detailed description of theoretical and practical aspects of pH electrodes is given by Bates [2], Westcott [3], and Buehler and Ingold [4].

B.   Sterilization and Sterility

Sterilization

Bioreactors are sterilized in situ or in an autoclave. The first method is less problematic for in situ sensors and will be discussed first. Only the tip of the combined pH electrode must withstand high temperatures and high pressure during the in situ sterilization. This is a solved problem if the coefficient of thermal expansion of pH and shaft glasses are matched. Compression tests revealed that a proper glass membrane will withstand pressures of up to 100 bar. With a liquid-type electrode, the pressurization of the electrolyte is necessary to avoid boiling.

Much more critical is the case of autoclaving where the full sensor is in contact with hot steam. Experience clearly showed that cables are water-permeable and lead to a much shorter lifetime of pH electrodes. Hence, autoclavable sensors must have a sterilizable electrode head as shown in Figure 1.

Sterility

There are at least two possibilities for infections caused by pH probes. The metallic probe or housing is sealed against the pH electrode and against the weld-in socket by O-rings. The qualities of the O-ring groove, the O-ring itself, and its position are of utmost importance. This fact is well known and is true for all in situ sensors. The second possibility concerns the infection caused by the flow of unsterile electrolyte through the porous diaphragm. The Ingold sensor uses a special diaphragm with pore sizes of close tolerances.

The mean pore diameter lies around 0.9 μm, which prevents infections by bacteria or fungi. Experience showed that infections caused by the diaphragm rarely occur in classical bioprocesses. Cultivation of genetically engineered organisms or of animal cells, however, are more critical and the prevention of viral infections is a must. Unfortunately, viruses may penetrate the diaphragm, and a sterile electrolyte is therefore needed. There are several possibilities designing appropriate pH sensors.

Sealed electrodes. Autoclaving of sealed electrodes leads to fully sterile electrodes. Sensors with a glass shaft cannot be completely filled with the electrolyte gel due to the small thermal expansion and poor flexibility of glass. Such electrodes must have a free inner space, which allows the expansion of the electrolyte. On the other hand, a fermentation medium may be forced under pressure into a reference electrode, causing potential shifts. Figure 2 shows a special Ingold sensor where the free inner space contains pressurized air to overcome the above drawback. The shelf life of such an electrode is, however, limited due to the decrease of the internal pressure with time.

Other electrodes use a plastic shaft which allows a complete filling with gelled electrolyte. Their shelf life is not critical. However, the use of plastic material is still questionable for the cultivation of animal or human cells. Plasticizers diffusing out of the sensor may influence the growth of cells.

Nonsealed electrodes. Autoclaving nonsealed electrodes is most easily accomplished by the insertion of the whole electrode into the bioreactor. Hence the opening for refilling the electrolyte is situated in the headspace of the bioreactor. A disadvantage of this idea is that the electrolyte can be contaminated by foam which is always present in the headspace. A more sophisticated but more complicated solution is a fully autoclavable, pressurized probe, recently developed by Ingold. The sterility of electrolyte and air is maintained by a sterile filter mounted inside the air valve. See Figure 3.

Figure 2   Construction of a prepressurized pH electrode: 1, auto-clavable screw cap; 2, pressurized air gap; 3, Ag/AgCl high temperature lead-off system; 4, gelled electrolyte.

C. Long-Term Stability and Accuracy

It is common practice to calibrate the pH sensors before mounting in the bioreactor. The subsequent sterilization should not change the calibration values. The cultivation periods may vary between hours and days or even weeks to months with certain cells. Therefore such cultivation of animal or human cells needs very stable sensors. Furthermore, as shown in Table 3, their optimal pH range is rather narrow. This example clearly shows that cell culture processes need more accurate sensors.

Table 3   pH Dependence of Growth

	Growth (optimal = 100%)	
	pH 7.2	pH 6.6
Melanoma cells	100%	0%
Escherichia coli	100%	95%

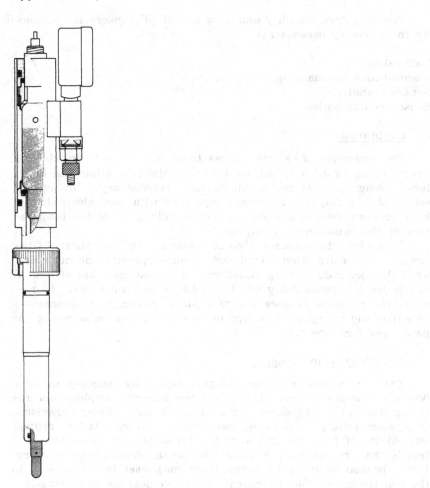

Figure 3 Fully sterilizable pH probe for small bioreactors.

The long-term stability and accuracy of pH sensors is influenced by the following parameters:

Calibration
Recalibration by sampling
Sensor stability
Sensor contamination

Calibration

The technique of calibration has to be correct and reproducible. Errors of up to 0.1 pH unit are easily obtained by either old buffers, wrong temperature, or wiping the electrode dry. Wiping dry may lead to a sluggish response, especially with used electrodes. Most accurate results are obtained when calibrating at the temperature of the subsequent cultivation.

Generally, the response time of sensors should be short. Otherwise, errors are often introduced because operators do not wait until the electrode reading stabilizes. The response time of new pH electrodes is unmeasurably short and below 30 s when old. In reality, the response is more sluggish due to electrical influences (electrostatic charging when wiping or washing and influence of capacitances from the cable, etc.).

Recalibration by Sampling

The on-line sensor is checked periodically by sampling and analyzing in the laboratory. The delay time between sampling and analyzing has to be very short if the sample contains living organisms. Otherwise, false pH values are measured. A filtrate is less critical, but release of CO_2 may still shift the pH value. A lesser-known fact is that the pH error is caused by an incorrect sample temperature. Manual or automatic temperature compensation relates only to the functioning of the electrodes. It is not possible to carry out any form of temperature compensation for the pH value of the sample solution. Examples are shown in Table 4. Cultivation at 40°C and measuring at 20°C would result in pH errors of up to 0.34 pH units (E. coli). Hence it is an absolute must to analyze samples at the temperature of cultivation.

Sensor Stability

pH glass electrode. The stability of modern pH glass electrodes is excellent. Zero point and slope vary within a narrow range, even after repeated sterilizations, as shown in Figure 4.

Reference electrode. The classic reference electrode with a liquid electrolyte and overpressure shows an excellent long-term

Table 4   Temperature Coefficients of the pH Value of
Different Solutions

Standard culture medium for	pH (20°C)	pH (30°C)	pH (40°C)
E. coli	6.98	6.80	6.64
Yeast	6.23	6.14	6.04
Fungi	5.79	5.75	5.71

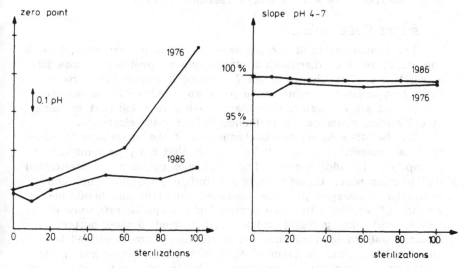

Figure 4   Stability of Ingold glass electrodes after repeated sterilizations at 121°C/20 min in distilled water.

stability. The overall shift after 100 sterilizations at 121°C for 20 min corresponds to less than 0.1 pH unit. The loss of KCl by flow and diffusion through the diaphragm does not influence the stability. In sealed electrodes, however, the loss of KCl by diffusion leads to a slow drifting of the potential and thus deterioration of the calibration values of the pH sensor. With a proper design and a large KCl reservoir, the sealed electrode may still show an acceptable lifetime. Knowing the total amount of dissolved KCl and the electrical resistance of the diaphragm, the lifetime can be estimated. The following rule of thumb is very useful:

$$L = VR \text{ month} \cdot mL^{-1} \cdot k\Omega^{-1} \quad \text{at } 25°C \quad (7.2)$$

where

L = lifetime in months, based on a potential shift corresponding to 0.1 pH units
V = volume of electrolyte or gel (mL)
R = electrical resistance of the diaphragm (kΩ) measured in 3 M̲ KCl

   The shifting of the potential is one aging property of sealed reference electrodes. The second phenomenon is an increase of the diaphragm potential with aging. The lower KCl concentration in the diaphragm is responsible for this fact [2]. The larger diaphragm potentials produce larger measuring errors, especially in dilute and acidic solutions, as well as longer response times.

Sensor Contamination

   The contamination of the pH-sensitive glass is very rare, though the literature often discusses contamination by proteins. Glass has a negative surface charge at pH 7. Hence it should bind proteins having an isoelectric point above pH 7 when cultivation is carried out around pH 7. Extensive experience has revealed that proteins of cultivation media do not influence the pH glass electrode.
   On the other hand, the contamination of the diaphragm by proteins is common. It is a well-known fact that many proteins are precipitated by adding salt. The diaphragm contains a concentrated KCl solution which therefore may precipitate proteins. Such contaminated diaphragms produce increased potentials and hence increased pH errors. In those applications, a special reference electrolyte containing glycerol and less KCl proved to work better. Another well-known contamination of the diaphragm is caused by silver sulfide. The reaction of AgCl with sulfur compounds in the broth causes such black diaphragms. Again, high potentials at the diaphragm may occur [4]. Other types of contaminations exist and depend on the composition of the medium.

Contamination of the diaphragm is one of the most important causes of faulty pH measurements in biotechnology. In principle, two methods exist to overcome this problem.

Special reference electrode. The formation of silver sulfide is partly eliminated by a bridge electrolyte and fully eliminated by the Ross electrode from Orion, by a Corning electrode having a barrier for silver ions [5], or by the Xerolyt electrode from Ingold. The last three sensors are not sterilizable. The Xerolyt electrode also eliminates other types of diaphragm problems, seen for example in anaerobic waste water treatment [6].

Diaphragm cleaning. Proteins are dissolved by a protease, and silver sulfide by an acidic solution of thiourea. Ingold developed retractable probes which allow the replacement or cleaning of electrodes without interrupting the sterility of the bioreactor. The newer version is showed in Figure 5. The sterilization of the cleaned or new sensor within the metallic probe is performed in two chambers. This method completely reduces the risk of an infection.

Several sources of errors have been described in Section IIC. Handling errors and the pH buffer solutions themselves are further sources of uncertainties. Therefore it is not realistic to reach an accuracy better than ±0.05 pH unit.

D. Flow Dependence

pH sensors do not suffer from a flow dependence of their signal with the exception of solutions with very low ionic strength. Problems arise if the specific conductivity of the solution falls below 10 µS/cm, which never happens in bioprocesses.

E. Interference

The pH glass electrode is characterized by excellent selectivity. Only sodium ions may interfere above pH 12. The reference electrode is less selective and may be influenced by all ions which form less soluble compounds with silver ions than with chloride. Such interferences are rare or even nonexistent in bioprocesses. An indirect interference is a ground-loop current flowing through the reference electrode. It will reduce silver chloride to silver and thus destroy the reference electrode. Ground loops are nonexistent when isolating amplifiers are used.

F. Conclusions

Table 5 summarizes the performance of present pH sensors in comparison to the requirements stated in Table 1. Table 5 shows that

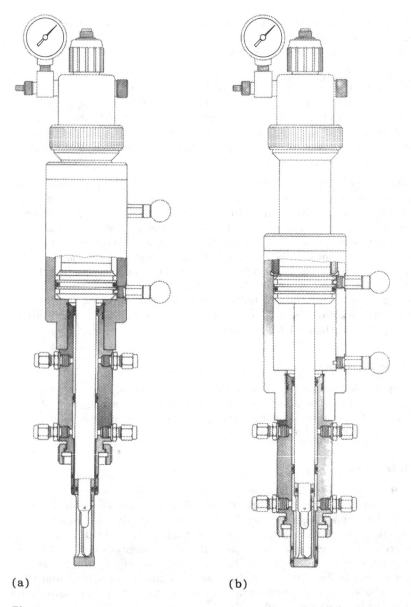

(a)   (b)

Figure 5 Construction of a retractable probe with two sterilization chambers: (a) measuring position; (b) retracted position.

Table 5 Status of pH Sensors Used in Bioprocesses

Requirements for in situ pH sensors	Suitable sensor and/or technique
Long-term stability	Ok for pH sensor with liquid electrolyte
	Retractable probes
	Special electrodes when having diaphragm contaminations
	Correct calibration and sampling procedures
Sterility	Fully sterilizable sealed electrodes, either with a plastic shaft or with a glass shaft and a prepressurized electrolyte
	Fully sterilizable and pressurizable probe
No interference	Amplifier with galvanic isolation
No flow dependence	No problem for biological media
Small size	Diameter down to 8 mm
No maintenance	Sealed sensors

still no ideal pH sensor exists, but it lists new sensors which eliminate some current limitations. There are good arguments that an ideal pH sensor will never exist. No protein precipitates at low KCl concentrations, but low diaphragm potentials need a highly concentrated reference electrolyte.

So far, only the characteristics of pH glass electrodes have been discussed. Several alternatives exist, which will be briefly mentioned. pH electrodes based on metal/metal oxide are well known. Most investigations deal with antimony and, recently, with iridium electrodes. These pH electrodes can be used at high temperatures, but they suffer from an unstable zero point, from the influence of oxygen and other oxidizing substances, and from interference by all substances which react with the metal oxide. Besides optical pH measurements many publications promote the ion-selective field-effect transistor (ISFET) pH electrodes. The ISFETs are of special interest because they can be produced in large quantities by using modern chip techniques. However, most publications describe the advantages and seldom address the other side. The drift is still far too large for an industrial application despite enormous improvements during the last few years. pH sensors using a plastic membrane have been developed. The drift is satisfactorily small. Interferences by lipophilic substances, and poor high-temperature stability prevent an application in the bioreactor.

III. DISSOLVED OXYGEN

A. Theoretical Aspects

Oxygen is another basic parameter in biological processes. Of particular importance are the effects of oxygen concentration or activity on the growth and metabolism of the organism. Below a certain oxygen value, metabolism changes. Therefore, most traditional aerobic bioprocesses demand a minimal amount of dissolved oxygen, which has to be controlled. Animal cells grow best when the dissolved oxygen is kept within limits close to those found in blood.

Essentially all of the dissolved-oxygen (DO) sensors operate on the principle of reduction of oxygen at the surface of a noble-metal electrode, the cathode. The problem of calibration and interferences by other solutes led to the introduction of the membrane-covered DO electrode, introduced by Clark. In its original design, both the Pt cathode and the Ag anode were contained in a single-electrode body containing the electrolyte solution. The entire tip of the sensor was covered with a gas-permeable membrane. Under a sufficient voltage between cathode and anode, the oxygen diffusing through the membrane is completely reduced at the cathode. This phenomenon can be observed from the current-voltage diagram, called a voltammogram. As shown in Figure 6, the current increases initially with an increase of the negative potential, followed by a plateau. In this plateau region, the reaction of oxygen at the cathode is so fast that the rate of reaction is limited by the diffusion of O_2 to the cathode. Going to more negative potentials, the current increases rapidly due to the reduction of water to hydrogen and hydroxide.

The current output of the electrode is proportional to the dissolved oxygen activity or partial pressure if a fixed potential in the plateau region is applied. This is the basic operating principle of the polarographic and galvanic DO probes. The galvanic DO probe differs from the polarographic type in that it does not require an external voltage source. Galvanic DO probes use a basic metal such as zinc or lead as the anode. See Figure 7. The principal differences between polarographic and galvanic oxygen electrodes concern only the anode, the electrolyte, and the polarization voltage.

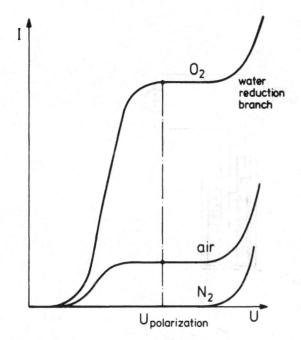

Figure 6 Voltammogram of a Clark DO sensor. Pt cathode, Ag/AgCl anode, neutral electrolyte.

The polarographic approach has the advantage that an optimal polarization voltage can be chosen. A diffusion barrier between the cathode and the Ag anode is necessary to avoid a plating of the cathode by silver diffusing to the cathode. The anode can be poisoned by hydrogen sulfide diffusing through the gas-permeable membrane. The corresponding shift of the anode potential may be large.

The galvanic approach has the advantage that the dissolved lead or zinc is not reduced at the cathode and the poisoning by hydrogen sulfide is less critical. On the other hand, the voltage may not be ideal. Furthermore, carbon dioxide may precipitate lead or zinc carbonate, which will reduce the active cathode surface and thus lower the electrode current. The storage time of galvanic sensors is limited because of the corrosion of lead or zinc by oxygen.

The current I of an ideal polarographic and galvanic DO sensor is derived from Fick's diffusion laws:

$$I = k \frac{(ADS)pO_2}{z} \qquad (7.3)$$

Bias voltage source

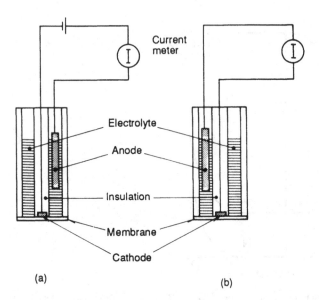

Current
meter

Electrolyte

Anode

Insulation

Membrane

Cathode

(a)   (b)

Figure 7 Principle of polarographic (a) and galvanic (b) oxygen
sensors.

in which k is a constant, A is the cathode surface and D, S, and z
are the diffusion coefficient, the solubility coefficient, and the
thickness, respectively, of the gas-permeable membrane, and pO_2 is
the oxygen partial pressure. The response time t of such a sensor
is

$$t = k' z^2 D^{-1} \qquad\qquad (7.4)$$

where k' is a constant. Therefore thin and highly permeable membranes are of advantage. On the other hand, the flow dependence
of the electrode current should be small, which is the case with low
electrode currents. DO sensors with a short response time and a
low flow dependence therefore have a thin membrane of low permeability and a small cathode. For a given membrane composition and
fixation, the stability of the electrode current grows with increasing
cathode surface. On the other hand, an increased cathode diameter
shortens the lifetime of DO sensors. These facts reveal that it is
almost impossible to develop an ideal DO sensor based on these
technologies.

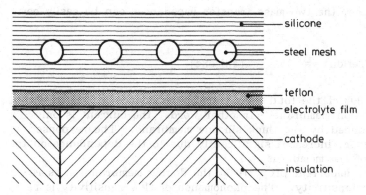

Figure 8 Reinforced double membrane of Haddad.

All common DO sensors have a relatively large cathode and a thin membrane in order to achieve good long-term stability and fast response. Those sensors, however, show a relatively large flow dependence, and therefore the measurements are affected by the intensity of agitation. A sensor made of a large cathode and a thin membrane is thus not an ideal sensor for biotechnological processes. Furthermore the thin membrane will not withstand large pressure differences.

The autoclavable polarographic DO sensor of Haddad [7] and Krebs and Haddad [8] has been commercialized by Instrumentation Laboratory and improved by Ingold. Haddad introduced the concept of a reinforced double membrane which proved to work well in bioreactors (Figure 8). The inner current-determining membrane is a 25-μm Teflon film. The outer membrane, approximately 150 μm thick, is a highly permeable silicone membrane reinforced by a thin stainless steel mesh.

Equation (7.5) gives the current for a DO sensor with a double membrane:

$$I = k \; \frac{A(pO_2)}{z/DS \; (Teflon) + z/DSF \; (silicone)} \tag{7.5}$$

The factor F (> 1) describes the influence of the nonaxial flux through the outer silicone membrane; z/DS may be called the mass transfer impedance of the oxygen gas. This relationship is valid for the membrane of a DO electrode as well as for the cell wall of an organism.

The permeability, DS, of silicone is roughly 90 times higher than that of Teflon. For the above-mentioned thickness of the dou-

ble membrane, the two mass transfer impedances can be easily compared:

$$\frac{z}{DS} \text{ (Teflon)} = 15 \frac{z}{DS} \text{ (silicone)}$$

More than 90% of the total mass transfer impedance is located within
the thin Teflon membrane. In other words, the current is only
slightly reduced by the thick silicone membrane. The response
time, however, increases significantly due to the dependence on the
thickness of the membrane.

Another important property of the double membrane is its reduced flow sensitivity. The phenomenon of flow sensitivity is related to the ease with which the sample can provide oxygen required
by the electrode. Its magnitude is easily determined by measuring
the sensor current in agitated and stagnant water. The degree of
flow dependence increases with the cathode area, membrane permeability, and, inversely, with the membrane thickness. These factors, which tend to produce a high output current and a short response time, also tend to increase flow sensitivity. The outer silicone layer of the double membrane acts as an oxygen reservoir or
stagnant layer with high oxygen-dissolving power. The thick silicone membrane therefore improves the supply of oxygen reduced by
the cathode in a stagnant medium. Table 6 compares DO sensors
having a single and a double membrane.

The first autoclavable oxygen electrode was a galvanic probe,
introduced by Johnson et al. in 1964 [9]. This electrode became
popular in biotechnology due to its ease of fabrication and relatively
large current output. Such sensors could directly be coupled to a
recorder. Several companies have further improved the original
Johnson probe (New Brunswick Scientific, Marubishi, Biolafitte,
etc.). At present the galvanic probe from New Brunswick (NBS)
and the polarographic electrode from Ingold are most popular. The
comparison of these two sensors will help to answer the old question: "Is the galvanic or the polarographic electrode the better approach?" Table 7 shows that the NBS and Ingold DO sensors differ
in many respects.

The Ingold sensor with the small cathode and the reinforced
double membrane is free from problems concerning pressure compensation, lifetime, flow dependence, and interference from air bubbles.
However, it needs several hours to stabilize when switching on the
polarization voltage. The first sterilization shifts the current by
5 to 10%. Further sterilizations increase the current by less than
4%. This increase is due to a deformation of the stretched silicone/
Teflon membrane. The calibration of DO sensors is always per-

Table 6 Performance of Polarographic DO Sensors with a 250-μm Pt Cathode and a Ag Anode at 25°C

	25 μm Teflon	25 μm Teflon + 150 μm silicone
Current, agitated water	54.6 nA	50.3 nA
Current, stagnant water	42.3 nA	48.9 nA
Flow dependence	22.5%	2.8%
Response time (95%) for nitrogen to air	20 s	50 s

formed after the sterilization. Hence, the above-mentioned change of the current does affect the accuracy.

These properties are hardly determined by the anode material, the electrolyte composition, and the polarization voltage. Hence the main differences of performance between the NBS and the Ingold probes do not originate from the different basic operating principles but from the different electrode designs.

The traditional galvanic and polarographic DO sensors are routinely used in bioreactors. However, it is worth mentioning other types of oxygen electrodes. The Mackereth probe is a galvanic DO electrode with a large tubular Ag cathode. It has a high current output which is stable for six months [10]. Sterilizable versions have been developed. Another interesting DO sensor has been patented by Leeds & Northrup [11]. It consists of a Pt cathode, a Pt anode, and a counterelectrode. The oxygen reduced at the cathode is formed again at the anode. Hence the net oxygen consumption is zero, which eliminates the flow dependence of the current. However, the response time is more sluggish, and the constancy of the thickness of the membrane and electrolyte film is of utmost importance.

Orbisphere introduced the concept of the guard ring which eliminates the contamination of the cathode by reduced silver chloride [12]. Oxygen electrodes based on the chip technique exist but have not yet gained importance.

Further theoretical and practical aspects of DO sensors are described by Lee and Tsao [13] and Hitchman [14].

B. Sterilization and Sterility

The Ingold sensor is completely closed. The membrane cartridge is enveloped with a silicone tube which allows the expansion of the

Table 7 Specifications of the New Brunswick and the Ingold
DO Probes

Design	NBS	Ingold
Cathode	Pt, 5 mm	Pt, 0.25 mm
Anode	Lead	Silver
Electrolyte	pH 13.5	KCl, pH 13
Polarization voltage	0 mV	−675 mV
Membrane	50 μm Teflon FEP	Double membrane (see text)
Diameter	14.5 mm	19 and 25 mm
Pressure compensation	Ventilation hole at the top of the probe	Reinforced membrane allows closed probe
Current (air, 25°C)	10 μA	40−80 nA

Performance		
Current (nitrogen)	Less than 1% of the current in air	
Response time (95%)	100 s	50 s
Conditioning	2 sterilizations	6 h
Polarization time	30 min	6 h
Drift at constant temperature	?	2%/week
Shift of current due to sterilization	?	2−10%
Flow dependence	High, velocity higher than 20 cm/s	2−4%
CO_2 interference	Yes	No
Lifetime of anode	6 months	5 years
Temperature compensation	No	Yes

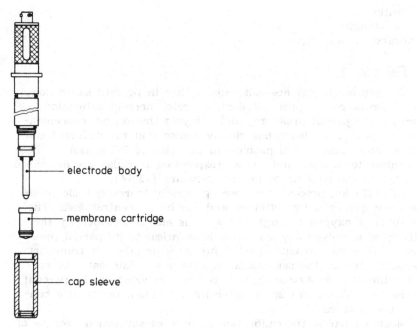

electrode body

membrane cartridge

cap sleeve

Figure 9   Sterilizable Ingold oxygen electrode.

electrolyte at high temperatures.   Flexible materials, such as silicone, show a high permeability for many volatile substances.   However, the electrolyte will not be contaminated by such substances
from the medium because of the perfect sealing of the cartridge by
the stainless steel cap sleeve.   The sensor has a sterilizable screw
cap and is therefore autoclavable (Figure 9).   The NBS probe is
also autoclavable but not completely closed.   Medium may enter the
ventilation hole and contaminate the electrolyte.   Problems of sterility do not exist if the O-ring positions are correct and some care
is taken for maintenance.

C.   Long-Term Stability and Accuracy

In many traditional cultivations the DO sensor was often used to
check whether enough oxygen was present.   The accuracy was of
minor importance.   Modern cell cultivation processes ask, however,
for a DO control within 10% of a fixed value.   Long-term stability
is therefore a parameter of increasing importance.   Stability and
accuracy of DO probes depend on

Calibration
Sensor stability
Membrane contamination

Calibration

Oxygen electrodes are calibrated either in percent saturation, partial pressure, or ppm. Biologists prefer percent saturation, chemists the partial pressure, and analysts the oxygen concentration in ppm. Experience has clearly shown that the different calibration units pose a real problem to the user of DO sensors. The calibration techniques and the corresponding formulas are described in detail in the manuals or in the literature [13,14].

All diffusion processes or more generally thermodynamic processes are governed by activities and not by concentrations. The transport of oxygen through cell walls is also determined by the activity of dissolved oxygen which is identical to its partial pressure. DO probes measure partial pressure or pO_2. If transport characteristics or thermodynamic aspects are of interest, the correct calibration or measuring unit is the pO_2 value or the percent saturation. These two units differ by a constant factor at a constant temperature.

Most popular is the calibration in percent saturation because of its simplicity. The calibration is performed after the sterilization in the well-aerated medium before inoculation. The medium must be completely saturated with air at the cultivation temperature. The saturation may last 10 to 30 min, depending on the bioreactor. It is very important to know that the 100% value depends on total pressure. Presently bioreactors seldom have a pressure indicator. If the internal pressure increases due to a clogged air filter, the DO probe may indicate a value greater than 100%. This is not an oversaturation. More oxygen is soluble at a higher total pressure which is identical to a higher pO_2 value indicated by the DO probe. The calibration in percent saturation is simple but confusing when changing pressures occur.

To calibrate in partial pressure, one needs formulas, tables, and a knowledge of the total pressure. On the other hand, its interpretation is clear and not confusing when working with different pressures.

The calibration in ppm is only justified if mass balances, etc., are calculated. The relation between oxygen partial pressure and oxygen concentration cO_2 follows from Henry's law.

$$pO_2 = cO_2 (H) \quad bar \cdot g^{-1} \cdot L \tag{7.6}$$

The Henry constant H depends on temperature, ionic strength, sugar concentration, organic solvents, etc. The Henry constant may vary during the cultivation which renders the continuous measurement of oxygen concentration unreliable. Schneider and Moser [15], however, developed a method for the determination of oxygen solubility during the cultivation. It is based on a joint analysis of gas and liquid phases. A fast-responding DO sensor is needed.

Sensor Stability

The stretched double membrane is responsible for the current increase after the first sterilization. Afterward the current remains reasonably constant as shown in Figure 10. Figure 10 does not give information about the electrode drift after the sterilization. The current decreases slightly during the first hours. The drift completely stops about two days after the sterilization. Depending on the time of calibration, the above drift is between 2 and 6%. Hence the DO sensor should be calibrated as late as possible or just before inoculation.

A further criterion for the stability is the electrolyte consumption, which is given by the diameter of the cathode, the diffusion resistance of the membrane, and of the type and volume of the electrolyte. At room temperature the Ingold DO probe can theoretically operate two years before changing the electrolyte.

Concerning the poisoning of the cathode or anode, no problems have been observed in aerobic bioprocesses.

Membrane Contamination

The silicone membrane has in most applications a sufficient chemical inertness. However, it may swell slightly in mediums containing large amounts of oil or hydrocarbons. The swelling alters the response characteristics of the sensor. The electrode output remains fairly constant due to the double membrane concept. The Teflon membrane is much less influenced by oil.

The other type of membrane contamination is the fouling by bacterial adhesion. The sensor current may drop to very low values if the adhering microbes are alive and consume oxygen. Such contaminations may be eliminated by a sudden increase of the impeller speed. A more elegant solution to this problem is the mounting of the DO probe where the turbulence is high. However, the pO_2 value may be too high at such a place, and therefore the sensor does not indicate useful values.

Different shapes of the electrode and modifications of the membrane have been evaluated in order to minimize microbial adhesion, so far without success. The only possibility is the use of a retractable probe, which enables the sensor to be cleaned. The present retractable probe 768-O_2 from Ingold is quite large and not suitable for small bioreactors.

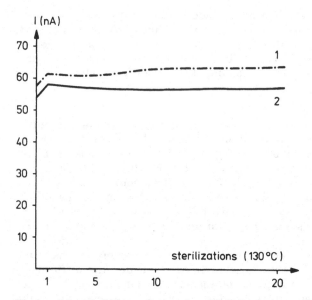

Figure 10 Stability of two Ingold DO sensors after repeated sterilizations at 121°C/20 min in distilled water.

Antifoam agents may adhere at the gas-permeable membrane and reduce the overall diffusion resistance: the lower the flow dependence of the DO sensor, the smaller the effect of antifoam agents.

D. Flow Dependence

The theory of flow dependence has been partly described in Section IIIA. A more detailed description of this topic is found in Refs. 13 and 14. The sensor with the silicone/Teflon double membrane is still the only industrial DO sensor with a small flow dependence of only 2—4%.

In highly viscous solutions the flow dependence is higher mainly due to the reduced diffusion coefficient of oxygen. The Stokes-Einstein relation indicates that the diffusion coefficient multiplied by the viscosity is constant. Ho et al. [16] have proved that this relation is valid for glucose-containing solutions.

E. Interference

Volatile substances are potentially interfering substances. Interferences come from gases which are reduced at the cathode. Examples are chlorine, bromine, iodine, some halogenated hydrocarbons, and nitrogen oxides, all of which are not present in bioprocesses.

Another well-known cathode interference is caused by carbon dioxide. It will decrease the pH value of the electrolyte and thus shift the voltammogram for the water reduction. This shift causes a higher sensor output. A buffered electrolyte eliminates the CO_2 interference so that the current in 100% CO_2 is less than 1% of the air current.

Anode contaminants are substances which alter the anode potential and therefore shift the voltammogram. The most important poison is hydrogen sulfide. Its influence is effectively reduced by a diffusion barrier between the gas-permeable membrane and the anode. An interesting approach to solve the H_2S interference is described by Hale [17]. He developed a DO sensor with a silver/silver sulfide anode. However, it requires an electrolyte which is unstable in contact with air.

F. Multiphase System. Effect of Air Bubbles

Air bubbles are not in a thermodynamic equilibrium with the liquid medium when living organisms are present. The pO_2 values may strongly differ, especially with a high biomass. It happens for example that the dissolved oxygen partial pressure is unmeasurably small despite 18% oxygen in the bubbles. What does the DO sensor show if gas bubbles stick on its membrane?

The simplest approach is an equal ratio of liquid and gas phases throughout the whole bioreactor. In other words, the same gas holdup exists also at the surface of the DO sensor. In this case the output of the sensor obeys Eq. (7.7):

$$pO_2(s) = a \cdot pO_2(l) + (1 - a) \cdot pO_2(g) \qquad (7.7)$$

where $pO_2(s)$ is the oxygen partial pressure indicated by the sensor; (l) and (g) mean liquid phase and gas phase, respectively; and a is the fraction of the liquid phase.

It follows from Eq. (7.7) that considerable errors arise if the gas holdup is high or if the $pO_2(l)$ is very small. Zimmer and Woelbing [18] proposed to correct the measured DO value according to Eq. (7.7). However, this procedure is an overcompensation of the gas bubble effect. The gas-permeable membrane is always covered by a liquid film which reduces the effect of the gas bubbles. Heinzle et al. [19] have evaluated the gas bubble effect. It turned out that modified membranes or modified geometries of the electrode tip are of minor importance. Bubble interference can be eliminated

Table 8 Status of the Sterilizable DO Sensors for Bioprocesses

Requirements for in situ DO sensors	Sensor characteristics
Long-term stability	Ok
	Mounting in a turbulent region to reduce membrane fouling
	Retractable probe for large reactors
Sterility	Fully sterilizable probe
No interference	Special electrolyte to eliminate CO_2 interference
	Mounting in a turbulent region and using a double membrane to reduce gas bubble interference
No flow dependence	Silicone/Teflon double membrane
Small size	Down to 12 mm

to a great extent by using a proper probe mounting in the reactor.
A highly turbulent region is best to avoid bubble impact and ad-
herence, but, as already mentioned, it may indicate an incorrectly
high pO_2 value. The bubble effect of DO probes is larger when a
vertical mounting, rather than an inclined mounting, is used.
Heinzle found air bubble effects of DO sensors (Figure 9) which
corresponded to 1—3% of the pO_2 value of the air bubble.
    Table 8 summarizes the performance of the sterilizable DO sen-
sors.

IV. REDOX POTENTIAL

A. Theoretical Aspects

In both living and nonliving nature, oxidation and reduction reactions are just as important as acid/base reactions. The oxidation
or reduction power of a solution is determined by the redox potential. It can also be considered as a measure of the ease with which
a substance either absorbs or releases electrons. The expression
"electron activity" is also common.
    The redox potential is measured with a redox electrode. It has
a configuration similar to a combined pH electrode. The pH glass

electrode is replaced by a platinum electrode. Sterilizable versions have a shape similar to the sensor of Figure 1.

Depending on the redox couple, the redox potential is pH dependent or independent:

$$Fe^{2+}/Fe^{3+}: \quad E = E_0 + s \log \frac{(Fe^{3+})}{(Fe^{2+})} \quad \text{pH independent}$$

$$O_2/OH^-: \quad E = E_0 + \frac{s}{4} \log \frac{(O_2)}{(OH^-)^4} \quad \text{pH dependent}$$

$$= E_0' + \frac{s}{4} \log(O_2) - s\,pH \quad\quad (7.8)$$

Clark has defined the rH value. It is a redox value which should be independent of the pH value. Since pH dependence is variable (depending on the redox couple) or even nonexistent, the original definition of rH is not fulfilled. Clark himself recognized this fact, and he wanted to retract the rH value. However, it was too late; this expression seemed to have an almost magical power of attraction, and the rH value is still in use! Nevertheless it is not the correct measuring unit to express the redox condition of a medium.

The magnitude of the redox potential depends also on the type of reference electrode. Either the reference electrode has to be clearly defined or the standard redox potentials have to be used. The latter are measured against the standard hydrogen electrode (SHE):

$$E_h = E + E_{ref}$$

where

E_h = redox potential against the SHE

E = redox potential against the reference electrode

E_{ref} = standard potential of the reference electrode + 200 mV at 35°C for a Ag/AgCl electrode in 3 $\underline{M}$ KCl

The main problem of redox measurements is the Pt electrode, which does not establish a reversible equilibrium with all redox couples present in the medium. In other words, only a few redox systems are detectable by the redox electrode. The quantitative description of the degree of reversibility between the Pt surface and the redox couple is the exchange current density I_0.

Table 9   Importance of the Exchange Current Density I_0

High I_0	Low I_0
Correct redox potential	Redox potential may be incorrect
Good reproducibility	Poor reproducibility
Rapid response	Sluggish response

The magnitude of I_0 is an indication of how easily the electrons migrate from the platinum to the redox system in the medium, and vice versa [20]. I_0 shows values ranging from 1 to $10^{-25} A/cm^2$. Quite generally platinum shows a much larger exchange current density than gold. Therefore platinum is the preferred metal for redox electrodes.

The exchange current density decreases linearly with decreasing concentration of the redox couple. Hence the redox measurement in dilute solutions is less satisfactory.

B.   Redox Potential in Microbial Culture

In most cases the redox potential is not a reliable measuring parameter despite the immense importance of redox reactions. This statement is based upon the following facts:

The cultivation medium is not in a redox equilibrium. Hence mixed potentials are measured.
The concentrations of the redox systems are generally low.
Most of the organic redox couples do not interact with the redox electrode due to an extremely small exchange current density.

Nevertheless the redox potential is monitored in bioprocesses. In some applications it is a valuable empirical parameter. Most of these investigations are based on the fundamental work of Jacob [21]. The redox system oxygen/water has a very low exchange current density. But in absence of other redox couples, the Pt electrode responds to dissolved oxygen quite well. Jacob [22] investigated bare Pt electrodes and Pt sensors covered with a gas-permeable membrane. These experiments revealed that the redox potential measured in fermentation media is mainly determined by the DO value.

The relationship between DO value and redox potential has been proven many times. However, the Pt electrode will never replace the DO probe due to poor accuracy. The Nernst equation for the oxygen/water redox couple (Eq. (7.8)) gives the theoretical pH and oxygen dependence of the redox potential:

60 mV for $\Delta pH = 1$
15 mV for $\Delta \log(O_2) = 1$

However, experiments normally give a dependence of only 30—35 mV per pH unit [24]. These facts indicate the interference of other redox couples present in the cultivation medium.

An advantage of the redox electrode is the fact that it shows an abrupt potential drop when changing from aerobic to anaerobic conditions. The Pt electrode responds especially well to dissolved hydrogen gas which may be present in anaerobic media. Heinzle and Lafferty [23] found an excellent correlation between $\log(pH_2/pO_2)$ and the redox potential.

Amino acid fermentations are normally performed under oxygen limitation, and quite often regular DO probes are not able to measure and accurately control DO values below 0.01 bar. Radjai et al. [24] successfully used redox electrodes to optimize these bioprocesses. The redox potential was controlled by manipulating the agitation speed.

Thompson and Gerson [25] tried to improve the reliability of redox potentials by adding a redox mediator to the medium. Neutral red was chosen because it does not influence bacterial growth. However, this approach is only reasonable if the redox mediator is in a thermodynamic equilibrium with the redox system of interest, for example with oxygen/water.

C. Characterization of Sterilizable Redox Electrodes

Redox electrodes resemble pH electrodes in many respects. The remarks concerning sterilization and sterility mentioned in Section IIB concerned mainly the reference electrode and are also valid for combined redox electrodes.

The long-term stability of the Pt electrode is sufficient. No recalibration is necessary. The mechanical cleaning or polishing of the Pt surface is advantageous from time to time. The response time is generally sluggish in biological mediums due to the low exchange current density. Hence it is not possible to measure reliable redox values in these samples. The continuous measurement of the redox potential is the only reliable technique. The problems with the reference electrode are less critical with redox than pH measurements because an error of 12 mV is tolerable for the redox potential but not for the pH value (12 mV = 0.2 pH unit).

Table 10   Status of Redox Electrodes Used in Bioprocesses

Requirements for in situ sensors	Sensor characteristics
Long-term stability	Ok
Sterility	Fully sterilizable and sealed electrodes
	Fully sterilizable probe
No interference	Big problem, depends on the question "what do we want to measure?"
No flow dependence	Problem of unknown magnitude
Small size	Diameter down to 8 mm
No maintenance	Sealed sensor
Easy interpretation of results	Big problem

Flow dependence and air bubble interference of the redox potential have to be taken into consideration. As soon as a reaction at the Pt electrode is limited by mass transport, a flow dependence of the signal must result. Such a reaction only occurs if coexisting redox couples are present which are not in thermodynamic equilibrium. Such a case is generally encountered in bioprocesses. However, the magnitude of flow dependence and bubble interference has not been carefully evaluated.

V.   DISSOLVED CARBON DIOXIDE

A.   Theoretical Aspects

The oxidation of carbohydrates to CO_2 and water is the basis for aerobic forms of life. The behavior of dissolved carbon dioxide in the cultivation media has, however, drawn little attention in the past. The major reason has been the lack of reliable sensors or measuring methods. Off-gas measurements with infrared gas analyzers or a mass spectrometer are reliable methods. Hence many researchers utilized the off-gas measurement to calculate the dissolved carbon dioxide concentration, assuming an equilibrium between the CO_2 partial pressures in the gas and liquid phases. The CO_2 concentration is calculated using Henry's law. This approach is quite debatable since the driving force for CO_2 desorption from

the liquid phase to the gas phase is proportional to the difference
in CO_2 partial pressures. Hence the pCO_2 value in the liquid medium is higher than in the off-gas. Otherwise no CO_2 desorption
is possible.

Yagi and Yoshida [26] measured dissolved carbon dioxide using
the so-called tubing method. A long coiled silicone rubber tube is
installed in the bioreactor. Nitrogen is forced through the tubing,
CO_2 permeates through the tube wall and enters the flowing nitrogen stream. The gas leaving the tube is then fed to an infrared
analyzer. The tubing method has been successfully used for the
detection of other dissolved volatile substances such as alcohols.
However, there are some complications in the use of the tubing
method because of the lack of equilibrium between CO_2 (medium)
and CO_2 (carrier):

Equipment which ensures a very constant and precise flow rate
   of nitrogen
Sensitivity to the contamination of the tube by oil or solids

A further disadvantage is the large volume of carrier gas required.
The second common technique for determining dissolved CO_2 involves the potentiometric carbon dioxide electrodes. These so-
called Severinghaus electrodes have been used in clinical chemistry
for nearly three decades. Such a pCO_2 sensor employs a gas-permeable silicone membrane which envelopes a bicarbonate solution,
forming a stagnant layer around a flat pH glass membrane. CO_2
gas diffuses across the silicone membrane until the CO_2 partial pressures are equal in the thin electrolyte film and in the medium. The
driving force is again the partial pressure and not the concentration. Hence it follows that the CO_2 sensor detects the partial pressure. In order to get short response times, the gas-permeable membrane must be thin and highly permeable. Among the homogeneous
membranes silicone is best. The microporous membranes are even
more permeable because CO_2 diffuses through the air-filled pores.
However, such membranes do not withstand high pressure differences. Liquid can penetrate the membranes above a certain pressure (intrusion point) which would terminate the functioning of a
pCO_2 sensor. The intrusion point may be dramatically reduced by
surfactants.

The electrolyte normally contains bicarbonate and chloride ions
and serves also as a reference electrolyte for the Ag/AgCl reference
electrode. A mesh or net is interposed between the gas-permeable
silicone membrane and the convex or flat glass membrane in order
to ensure constant thickness of the electrolyte film. The shape of
the pH glass membrane and the thickness of the electrolyte film are
of utmost importance in minimizing the edge effects related to the

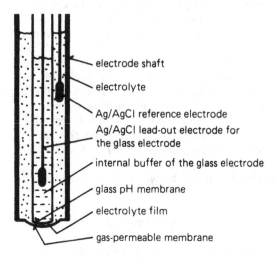

electrode shaft

electrolyte

Ag/AgCl reference electrode

Ag/AgCl lead-out electrode for
the glass electrode

internal buffer of the glass electrode

glass pH membrane

electrolyte film

gas-permeable membrane

Figure 11 Prinicipal design of a pCO_2 electrode.

electrolyte reservoir. Experiments have shown that the response
time strongly increases when the thickness of the electrolyte film is
increased.

The CO_2 diffusing through the silicone membrane reacts with
water. The hydration of CO_2 is a fairly slow process. The reaction half-time is about 20 s at 25°C.

$$CO_2 + H_2O \rightleftharpoons HCO_3^- + H^+$$

The pH value of the electrolyte film is governed by the following
equation:

$$\frac{(H^+) \cdot (HCO_3^-)}{pCO_2} = K \tag{7.9a}$$

The potential U of the combined pH sensor shown in Figure 11
obeys Equation (7.9b):

$$U = E_g + s \log(H^+) - E_r + s \log(Cl^-) \tag{7.9b}$$

where E_g is the standard potential of the glass electrode, E_r is the
standard potential of the reference electrode, (Cl^-) is the chloride

activity of the reference electrolyte, and s is the slope. Combining Eqs. (7.9a) and (7.9b) gives

$$U = E_0 + s \log pCO_2 - s \log(HCO_3^-) + s \log(Cl^-) \qquad (7.10)$$

where all constants are included in E_0. It follows from Eq. (7.10) that the Severinghaus CO_2 electrode has a stable zero point if the bicarbonate concentration in the electrolyte film and the chloride concentration in the reference electrolyte are constant.

Equation (7.10) shows that the CO_2 electrode cannot be calibrated with nitrogen gas. The potential is not clearly defined. Hence, a zero-point calibration with zero CO_2 is not possible. The gas-permeable membrane is also permeable to water vapor. Any transport of water through the silicone membrane causes a shift in the bicarbonate concentration within the thin electrolyte film and therefore a shift of the calibration values. If the sensor is properly designed, it is possible for the bicarbonate and the chloride concentrations to change by the same factor. Hence the calibration values remain stable despite the water transport through the silicone membrane. A reference electrode positioned near the gas-permeable membrane is more susceptible to poisoning, for example by hydrogen sulfide.

In 1980 Puhar et al. [27] described the first sterilizable carbon dioxide electrode which was based on the principal ideas of the sterilizable oxygen probe. The construction of the pCO_2 probe is shown in Figure 12. The probe consists of the protecting tube 1 with an interchangeable membrane cartridge 4, an electrolyte chamber 7 with supply lines for calibration 6 and the electrode shaft 2 with the combined pH electrode 3. The electrode 3 and the shaft 2 can axially be moved by means of a knob 5. This enables the two functions "calibration" and "measuring." Therefore the pH electrode can be moved back and a buffer solution corresponding to a certain pCO_2 value can be introduced between the silicone membrane and the glass electrode. After the buffer is replaced with electrolyte, the glass electrode can mechanically be moved forward until it again contacts the silicone membrane (measuring position). The silicone membrane which closes the cartridge 4 is reinforced by a stainless steel mesh. A nylon net guarantees a constant thickness of the electrolyte film. The reinforced membrane ensures constant tip geometries. In other words, the edge effects remain within narrow limits. Edge effects are caused by CO_2 diffusion from the electrolyte reservoir to the thin film, and vice versa. Step changes from high to low CO_2 levels lead to a sluggish response due to the above edge effect.

The pCO_2 probe is advantageously calibrated in CO_2 partial pressures. Other measuring units such as percent CO_2 saturation are feasible and are discussed in Section III for the oxygen probe.

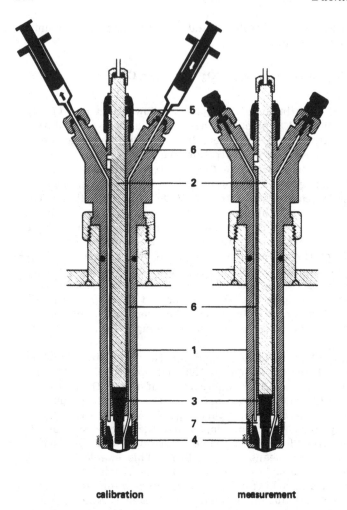

calibration   measurement

Figure 12   Sterilizable carbon dioxide sensor. (For 1–7, see text.)

Shoda and Ishikawa [29] described an alternative sterilizable
CO_2 sensor, which has a simpler construction than the Ingold sensor.   The inner combined pH electrode is inserted after the sterilization of the membrane holder.   A nonreinforced silicone membrane
closes the holder.   Therefore a dummy rod has to be installed into
the holder during the in situ sterilization.   This is in practice also
done with the Ingold pCO_2 probe, because the lifetime of the inner
pH electrode is increased considerably.

The availability of sterilizable carbon dioxide sensors has improved our understanding pertaining to the effect of dissolved CO_2. Most of these investigations are summarized in Ho and Shanahan [30]. At values of pCO_2 below 50 mbar, the growth of most important industrial microorganisms is either relatively unaffected or improved by the presence of dissolved CO_2. Above 100 to 200 mbar CO_2, however, cell growth and general metabolic activities are both adversely affected.

Smith and Ho [31] studied the behavior of dissolved carbon dioxide in penicillin fermentations. They demonstrated that the partial pressures of CO_2 in the off-gas and in the medium were not in equilibrium. The difference was greatest during and after the exponential growth phase. They measured values such as 20 mbar for off-gas and 70 mbar for dissolved CO_2. In such cases the controlling of both off-gas and dissolved CO_2 is advantageous. Lee et al. [32] examined the influence of dissolved carbon dioxide on the production of the antibiotic sisomicin and found that the higher the aeration, the lower the pCO_2 value and the higher the concentration of sisomicin.

Leist et al. [33] investigated the cultivation of a human melanoma cell line with and without process control. Without control the pCO_2 value increased to over 300 mbar. The viability of the cell line dropped to less than 50% above a pCO_2 value of 250 mbar. Suspension cultures with controlled pH and DO level also led to optimal dissolved CO_2 values between 50 and 100 mbar.

B. Sterilization and Sterility

The pCO_2 probes are completely closed. However, the present carbon dioxide sensors are not autoclavable.

C. Long-Term Stability and Accuracy

Cell cultivation processes especially require a controlled pCO_2 value. A reasonable range would be a CO_2 partial pressure of 75 ±25 mbar. This range corresponds to ±7.5 mV. Stability and accuracy depend on calibration, sensor stability, and membrane contamination.

Calibration

Two means of in situ calibration are feasible:

1. Calibration of the internal pH electrode by injecting buffer solutions
2. Calibration in the fermentor by saturating the medium with CO_2 gases of fixed composition before inoculation

Advantages of the first technique are

Fast response
Possibility for recalibration during the process

The disadvantages are

Increased uncertainty to ±10%.
The flushing of buffer and electrolyte reduces the lifetime of the
reference electrode due to the stripping off of silver chloride.
Probe diameter of 25 mm.
Unsuitable for determining effects of the gas-permeable membrane.

Spinnler et al. [28] stressed the last point. It is correct that alterations of the membranes cannot be detected. However, deposits
on the silicone membrane such as a film of antifoam slow down the
response characteristics, but the end value is still correct.
    Method 2 is more precise. An accuracy of 2% is feasible. How-
ever, the whole bioreactor has to be saturated by the CO_2 calibration gas. Such a method is only realistic with small fermentors.
The total pressure in the bioreactor has to be exactly known.

Sensor Stability

    The sensor stability is influenced by the glass-reference elec-
trodes as well as by the electrolyte. Equation (7.10) clearly shows
the importance of a constant electrolyte composition. The probe
shown in Figure 12 allows the introduction of fresh electrolyte which
is necessary with long cultivation durations. Such a replacement
does, however, strip off some silver chloride of the reference electrode, which shortens its lifetime. The glass electrode has a good
stability. Figure 13 shows the stability of the CO_2 sensor after
repeated sterilizations.
    The stability of a carbon dioxide sensor is inferior to a pH
electrode due to more sources of errors.

Membrane Contamination

    The membrane contamination of the carbon dioxide sensor is less
critical than with the DO probe. This fact is due to the potentiometric measuring principle. A layer of oil or antifoam slows down
the response of the CO_2 sensor, but the end value is still correct.
    A contamination by living organisms, however, leads to incor-
rect results. In this respect the carbon dioxide and oxygen sensors behave similarly. The different ways to reduce this effect are
described in Section IIIC.

pCO$_2$ (mbar)

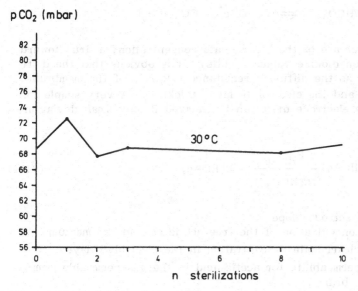

Figure 13   Stability of the pCO$_2$ probe.

D. Flow Dependence

A flow dependence is not observed.   Hence the carbon dioxide sensor is well suited for the cultivation of cells in non- or slightly stirred cultivation vessels.

E. Interferences

The operating principle of the Severinghaus electrode is based on the acidic behavior of the CO$_2$ gas diffusing into the sensor. Therefore all volatile acidic and alkaline substances are potential interferences.

A potential alkaline-interfering substance is ammonia, which is used in some bioprocesses to control the pH value.   At the same time ammonia serves as the nitrogen source.   However, most bioprocesses are conducted between pH 4 and pH 7.5 so that hardly any free ammonia is present.

Similarly in a neutral medium, most of the acetic acid is dissociated.   Neither ammonium nor acetate ions can interfere.   However, at lower pH values more acetate is transformed into its corresponding acid, reaching 50% at pH 4.7.   When diffusing through the gas-permeable membrane, acetic acid titrates the bicarbonate in the electrolyte layer according to the reaction

$$HOAc + HCO_3^- \rightleftharpoons OAc^- + CO_2 + H_2O$$

Due to a decrease of the bicarbonate concentration, a drift toward higher carbon dioxide values results. It is obvious that the drift gets smaller as the diffusion resistance z/P_{HOAc} of the membrane gets higher and the electrolyte layer thickens. A very simple formula for the electrode drift can be derived if only small deviations are allowed.

$$\frac{dE}{dt} = s(HOAc) \frac{P_{HOAc}}{za(HCO_3^-)} = k(HOAc) \qquad (7.11)$$

where s = electrode slope
(HOAc) = concentration of the free acetic acid in the medium
(HCO_3^-) = bicarbonate concentration in the electrolyte layer
P_{HOAc} = permeability for acetic acid in the gas-permeable membrane
z = thickness of the membrane
a = thickness of the electrolyte layer

Hence, the drift dE/dt is proportional to the concentration of acetic acid in the cultivation medium. The constant k is a function of the CO_2 sensor. The drawback of using a thick membrane and a thick electrolyte layer to reduce the HOAc interference is a sluggish response. Similar to the oxygen electrode a compromise has to be found to optimize the performance of the carbon dioxide sensor.

The sensor shown in Figure 12 allows a periodic replacement of the contaminated electrolyte. If the deviation of the true value should not exceed 10%, a replacement of the electrolyte is adequate after

> one year if the medium has a pH value of 7.0
  two days if the medium has a pH value of 4.7

The above results follow from Eq. (7.11) and are valid at 30°C and if the medium contains a total acetate concentration of 1 mmol/L.

F. Effect of Air Bubbles

The general aspects of the interference of air bubbles which have been described for the DO sensor are also valid for the carbon dioxide electrode. The experiments of Heinzle et al. [19] revealed that the bubble effect for the DO electrode was about 2% of the pO_2

level within the air bubble. Hence the effect is only noticeable at very low concentrations of dissolved oxygen.

Such a high difference between the partial pressures in the air bubbles and in the medium never exists for carbon dioxide. The maximum difference between these pCO_2 values may be a factor of 4. By the results of Heinzle, the dissolved CO_2 is lowered by less than 1% by the interference of air bubbles.

G. Maintenance and Size

The sterilizable carbon dioxide sensors are not as easily handled as pH or DO sensors. The calibration technique is especially time-consuming and needs practice and care.

Obviously the calibration method of injecting buffers led to a greater diameter of the sensor. It is hardly possible to decrease the present diameter of 25 mm. Sensors according to the ideas of Shoda have a diameter of 16 mm, which may be further reduced.

Table 11   Status of Sterilizable Carbon Dioxide Sensors

Requirements	Sensor characteristics
Long-term stability	Depends on the composition of the medium (interfering substances)
	Ok if the electrolyte is replaced periodically
	Mounting in a turbulent region to reduce membrane fouling
Sterility	Closed sensor
	No autoclavable version available
No interference	Special sensor to reduce the interference by acidic and alkaline substances
	Sensor which allows the replacement of the electrolyte
No flow dependence	Ok
Small size	Not available
No maintenance	Not available

VI. ION-SELECTIVE ELECTRODES (ISE)

Glass research has not only improved the specifications of pH electrodes, but has also led to cation-selective electrodes. The extensive work of Eisenman et al. [34] was of special importance. He developed sodium electrodes with a reasonable selectivity toward other cations. Glass electrodes for measuring potassium and ammonium have also been fabricated, but they show a poor selectivity. Hence they have not gained importance in analytical chemistry.

The sodium glass electrode is still widely used in clinical chemistry and in the control of power stations. Its configuration equals that of a pH electrode. The pH glass membrane is replaced by a sodium glass membrane. However, its selectivity is limited, which forbids universal application. The selectivity of potentiometric electrodes is best described by the Nicolsky equation

$$E = E_0 + \frac{s}{z} \log(a_i + K_{ij}a_j) \tag{7.12}$$

where a_i = activity of the measured ion

j = interfering ion

z = charge of the ions i and j

s = slope

K_{ij} = selectivity factor

K_{ij} depends on the concentration and is therefore not a constant.
The main interfering ions for the pNa sensor are silver and protons.
The following selectivity factors are valid for the sodium glass NAS 11-18 from Eisenman:

$$K_{Na/Ag} = 500$$

$$K_{Na/H} = 50$$

Hence the pNa electrode is more selective toward silver and protons, and the concentrations of these ions have to be very low. The following rule is important to keep the H^+ interference below 5%:

$$pH \geq pNa + 3 \tag{7.13}$$

where $pNa = -\log(Na^+)$. It follows from Eq. (7.13) that

pH 7: $(Na^+) \geq 10^{-4}$ $\underline{M}$

pH 4: $(Na^+) \geq 10^{-1}$ $\underline{M}$

Sodium measurements can be performed only in neutral and basic solutions. Experiments showed that the sodium electrode is sterilizable. Zero point and slope remained constant, but response time increased. So far the continuous measurement of the sodium level in the medium has drawn little attention.

A second sterilizable ISE is the H_2S sensor developed by Frevert [35]. Originally it was designed for monitoring hydrogen sulfide in water and sediment-water systems. Later it was found to be sterilizable and showed a good long-term stability. It will gain importance in anaerobic waste water treatment because the growth of many organisms is inhibited by hydrogen sulfide. Another example is sulfate-reducing bacteria, where optimal growth is only attainable if the H_2S concentration is kept below a certain level [36].

The H_2S sensor consists of a pH glass electrode and a silver/silver sulfide electrode. Its big advantage is the lack of a reference electrode. Hence the H_2S sensor is free from problems concerning sterility and maintenance.

The combination of the Nernst equations of the pH electrode and the sulfide electrode together with the dissociation constant for H_2S leads to Eq. (7.14):

$$U = U_0 - 0.5s \; \log(H_2S) \qquad (7.14)$$

Between 1960 and 1970 a whole new range of ISEs has been developed. This new field of sensors continues to grow at an increasing rate. The review article of Meyerhoff and Fraticelli [37] contains more than 500 references which cover a period of only two years. The role of ISEs in microbial process control has been summarized by Clarke et al. [38]. He mentioned over 300 publications. These solid-state ISEs, liquid and polymer membrane ISEs, and gas sensors can be combined with enzymes, tissues, or immobilized cells, which opens an exciting field for new applications. However, all these sensors are not steam-sterilizable and not yet applicable as real in situ sensors. Therefore the on-line process analyzers with ISEs are based on continuous sterile sampling of the cultivation medium. Quite generally, the long-term stability and selectivity of these electrodes are often moderate. On the other hand, the analyzer outside the bioreactor allows a periodic recalibration as well as preconditioning of the sample to eliminate interferences.

It is beyond the scope of this chapter to mention the many applications of ISEs for controlling bioprocesses. Furthermore these applications have not yet reached a stage of reliability. One example is the ammonia gas sensor. It has a high selectivity and has been used for the on-line measurement of ammonium in media [39, 40]. A frequent recalibration is necessary to compensate for zero-point drift.

In conclusion the future of ISEs seems for the time being outside the bioreactor. Great efforts are necessary to develop reliable and robust analyzers requiring minimum maintenance.

REFERENCES

1. Ingold, W., Dechema-Monographien, 43, 153 (1961).
2. Bates, R. G., Determination of pH, Wiley, New York (1973).
3. Westcott, C. C., pH Measurements, Academic Press, New York (1978).
4. Buehler, H. and Ingold, W., Process Biochem., 11, 19 (1976).
5. Brezinski, D. P., Anal. Chim. Acta, 134, 246 (1982).
6. Buehler, H. and Baumann, R., Labmate, 11, 34 (1986).
7. Haddad, I. A., U.S. Patent 3,718,562 (1973).
8. Krebs, W. M. and Haddad, I. A., Develop. Ind. Microbiol., 13, 113 (1972).
9. Johnson, M. J., Borkowski, J., and Engblom, C., Biotech. Bioeng., 6, 457 (1964).
10. Mackereth, F. J. H., J. Sci. Instrum., 41, 38 (1964).
11. Connery, J. G., Muly, E. C., and Taylor, R. M., U.S. Patent 4,076,596 (1978).
12. Hale, J. M., U.K. Patent Application GB2,013,895A (1979).
13. Lee, Y. H. and Tsao, G. T., Adv. Biochem. Eng., 13, 35 (1979).
14. Hitchman, M. L., Measurement of Dissolved Oxygen, Wiley, New York (1978).
15. Schneider, H. and Moser, A., Biotech. Lett., 6, 295 (1984).
16. Ho, C. S., Ju, L., and Ho, C., Biotech. Bioeng., 28, 1086 (1986).
17. Hale, J. M., in Polarographic Oxygen Sensors (E. Gnaiger and H. Forstner,eds.), Springer-Verlag, New York (1983), Chap. 1.6.
18. Zimmer, G. and Woelbing, M., Lebensmittelindustrie, 25, 489 (1978).
19. Heinzle, E., Moes, J., Griot, M., Sandmeier, E., Dunn, I. J., and Bucher, R., Annals N Y Acad. Sci., 469, 178 (1986).
20. Buehler, H. and Galster, H., Redox Measurement, Ingold Messtechnik, CH-8902 Urdorf.

21. Jacob, H. E., <u>Methodes Microbiol.</u>, <u>6B</u>, 91 (1972).
22. Jacob, H. E., <u>Biotech. Bioeng. Symp.</u>, <u>4</u>, 781 (1974).
23. Heinzle, E. and Lafferty, R. M., <u>European J. Appl. Microbiol.</u>
    <u>Biotechnol.</u>, <u>11</u>, 17 (1980).
24. Radjai, M. K., Hatch, R. T., and Cadman, T. W., <u>Biotech.</u>
    <u>Bioeng. Symp.</u>, <u>14</u>, 657 (1984).
25. Thompson, B. G. and Gerson, D. F., <u>Biotech. Bioeng.</u>, <u>27</u>,
    1512 (1985).
26. Yagi, H. and Yoshida, F., <u>Biotech. Bioeng.</u>, <u>19</u>, 801 (1977).
27. Puhar, E., Einsele, A., Buehler, H., and Ingold, W., <u>Bio-</u>
    <u>tech. Bioeng.</u>, <u>22</u>, 2411 (1980).
28. Spinnler, H. E., Bouillanne, C., Desmazeaud, M. J., and
    Corrieu, G., <u>Appl. Microbiol. Biotechnol.</u>, <u>25</u>, 464 (1987).
29. Shoda, M. and Ishikawa, Y., <u>Biotech. Bioeng.</u>, <u>23</u>, 461 (1981).
30. Ho, C. S. and Shanahan, J. F., <u>CRC Crit. Rev. Biotechnol.</u>,
    <u>4</u>, 185 (1986).
31. Smith, M. D. and Ho, C. S., <u>Chem. Eng. Commun.</u>, <u>37</u>, 21
    (1985).
32. Lee, J. L., Gil, G. H., Cho, Y. J., and Yoo, M. Y., <u>Kor. J.</u>
    <u>Appl. Microbiol. Bioeng.</u>, <u>14</u>, 355 (1986).
33. Leist, C., Meyer, H. P., and Fiechter, A., <u>J. Biotechnol.</u>,
    <u>4</u>, 235 (1986).
34. Eisenman, G., Bates, R., Mattock, G., and Friedman, S. M.,
    <u>The Glass Electrode.</u> Wiley, New York (1965).
35. Frevert, T., <u>Schweiz. Z. Hydrol.</u>, <u>42</u>, 255 (1980).
36. Cypionka, H., <u>J. Microbiol. Methods,</u> <u>5</u>, 1 1986.
37. Meyerhoff, M. E. and Fraticelli, Y. M., <u>Anal. Chem.</u>, <u>54</u>, 27R
    (1982).
38. Clarke, D. J., Kell, D. B., Morris, J. G., and Burns, A.,
    <u>Ion-Selective Electrode Rev.</u>, <u>4</u>, 75 (1982).
39. Thompson, B. G., Kole, M., and Gerson, D. F., <u>Biotech.</u>
    <u>Bioeng.,</u> <u>27</u>, 818 (1985).
40. Kuhlmann, W., Meyer, H. D., Bellgardt, K. H., and
    Schuegerl, K., <u>J. Biotech.</u>, <u>1</u>, 171 (1984).

8

Electrochemical Biosensors for Bioprocess Control

ROBERT J. GEISE and ALEXANDER M. YACYNYCH   *Rutgers, The State University of New Jersey, New Brunswick, New Jersey*

I. INTRODUCTION

As the field of electrochemical biosensors expands, there may be many applications for monitoring bioprocesses. Although not currently used to any great extent as a conventional sensor in bioprocess control, the interest in potential applications for biosensors is being driven by the need to control and optimize bioprocesses. In particular, biosensors hold promise for monitoring and optimizing production of "high-value added" products, such as pharmaceuticals. In fact, a consulting firm has done a marketing study which predicts that the demand for biosensors in process control will exceed that of the clinical market in the future.

Biosensors for bioprocess control should be designed to monitor substrate uptake, growth rate and production and provide a feedback system triggered by sensor response. The major limitations of biosensors are instability of the biocomponent, the need for fast, reliable calibration, and the requirement, in many processes, for sterilization of the sensor. The ideal sensing system for bioprocess control would be in situ or on-line, and continuous, with data obtained in real time, as opposed to conventional, time-consuming offline methods.

A promising development uses flow injection analysis (FIA), where a very small sample (e.g., 5 µL) of the reaction broth is automatically and periodically fed into a flowing stream and monitored with a biosensor. What results is a system that can be calibrated as needed, and can also be automated to trigger a feedback mechanism which adjusts levels of substrates in the reaction broth. This type of system and others applicable for monitoring bioprocesses, and their inherent disadvantages, will be discussed in later sections.

A. Types of Electrochemical Sensors

There are three types of electrochemical sensors categorized by the
mode of measurement: amperometric, potentiometric, and conducto-
metric. Conductrometric sensors are rare because they are nonselective and have poor signal-to-noise ratios. However, recent developments of these sensors in analytical biochemistry have been reviewed [1]. Potentiometric sensors utilize an ion-selective electrode
or gas-sensing electrode to monitor a certain ion or gas. These
sensors are selective and have wide response ranges, but have the
disadvantage of slower response times. An amperometric biosensor
measures the current generated by an electrochemical reaction at a
fixed potential. In the amperometric mode, the reaction of interest
must generate a change in the concentration of a species that is either oxidizable or reducible. Amperometric-based sensors have good
sensitivity and linear range, however, they lack the selectivity of
ion-selective electrodes, because they respond to any species that
is electroactive at the working potential.

B. Construction of Electrochemical Biosensors

An electrochemical biosensor consists of two parts: a biochemical
transducer and a physical transducer. The former can consist of a
variety of biocomponents, such as antibodies, receptors, enzymes,
tissues, or microbes. These are immobilized onto the physical
transducer (the electrode surface), usually platinum or a form of
carbon (amperometric), or ion-selective or gas-sensing electrodes
(potentiometric). The specificity of the biosensor can be controlled,
to a large degree, through the choice of the biochemical transducer.
The working sensor allows measurement of the electroactive products
or reactants of the biologically mediated reaction. As a result a
substrate, such as glucose, can be quantitatively monitored because
its oxidation by glucose oxidase produces the electroactive species,
hydrogen peroxide. The response to hydrogen peroxide is proportional to the concentration of glucose. A wide range of biocomponents permits the determination of a wide variety of biologically significant compounds.

C. Advantages of Electrochemical Biosensors

There are several key advantages to electrochemical biosensors that
make them very useful in bioprocess control [2]. First, electrochemical biosensors are inexpensive, relatively simple in theory and
operation and are adaptable to process control instrumentation. The
biocomponent chosen results in a biosensor that is very selective.
Biosensors can operate over a wide concentration range, encompass-

ing the range of interest for conventional bioprocesses [3,4], as well as micromolar [5] and even nanomolar [6,7] levels. Also, a colored or turbid sample that would present problems in spectroscopic analyses would not interfere in an electrochemical analysis [3]. There are several recent reviews on the construction and application of biosensors in general [8—15], biosensors applied to bioprocesses [2,3,16—20], and in particular, microbial sensors for use in bioprocesses [21,22].

II. REAL-TIME ANALYSIS AND SAMPLE HANDLING

Figure 1 depicts four possible configurations for monitoring bioprocesses with biosensors: (a) in situ; (b) continuous-flow loop system; (c) flow-injection analysis; (d) off-line. In-situ analysis involves placing the electrochemical biosensor directly in the reaction broth, as is typically done for pH measurements. If the response triggers a feedback control to adjust broth composition, this control mode is known as a closed-loop feedback system. There are inherent problems in placing any probe directly into a complex solution without any prior sample pretreatment. For electrochemical biosensors, the major problem with an in situ determination is fouling of the electrode surface by biomacromolecules, which results in poor electrochemical behavior of the sensor. This problem and methods used to overcome it are discussed later. In a continuous-flow loop system, the reaction broth flows through an external loop containing the biosensor, then back into the reaction broth. The sensor response can be continuously monitored and adjustments made automatically by a control system. The problems with this method are similar to the in situ probe arrangement, since again the sensor is constantly exposed to the broth. Also, this constant exposure necessitates sterilization of the sensor and the flow loop, because the broth is returned to the reaction vessel. The advantage of this sampling system, however, is that it provides a continuous, real-time broth composition analysis.

Another technology applicable for bioprocess monitoring is flow injection analysis (FIA). FIA methods have the advantage of exposing the sensor to the broth only periodically and since the sample analyzed by the biosensor does not return to the reaction vessel, sterilization of the sensor is not necessary. The disadvantage is that the monitoring is not continuous, but periodic. A detail of an FIA system suited for process monitoring is shown in Figure 2. There are three valve positions, corresponding to three solutions: (a) buffer solution, (b) calibrant, and (c) sample. Also, note that a multiple biosensor array in parallel can be used. For actual anal-

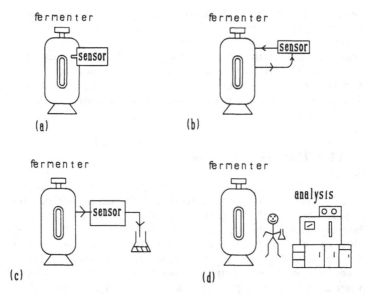

Figure 1 Schematic of four configurations used in the monitoring of bioprocesses with biosensors. (a) in situ; (b) continuous-flow loop; (c) flow-injection analysis; (d) off-line.

ysis of a broth sample, one switches from the flow of buffer to the sensor(s) to allowing the calibrant solution to flow to the sensor(s). Calibration can be automated or done manually. The valves are then switched again to allow a small sample of broth (5 μL) to be analyzed by the sensor(s).

Off-line methods involve manually removing a sample from the process and analyzing it. A common example is gas chromatography/ mass spectrometry (GC-MS) for the analysis. There are inherent problems with off-line monitoring. The major problem is that the sample may not represent the batch because of the time lag inherent in the analysis. As a result conventional off-line monitoring methods are not compatible with automatic feedback control which requires that the state of the process is known (i.e., real-time analysis) and that any adjustments are made as a result of changes in the conditions of the batch. Also, frequent sampling leads to a greater risk of contamination.

Enfors and Cleland have presented a discussion of the advantages and disadvantages of both external and in situ sensors [23]. For electrochemical biosensors, in situ use precludes any sample pretreatment, making it impossible to "clean up" the sample, which can lead to fouling of the electrode surface by absorption of bio-

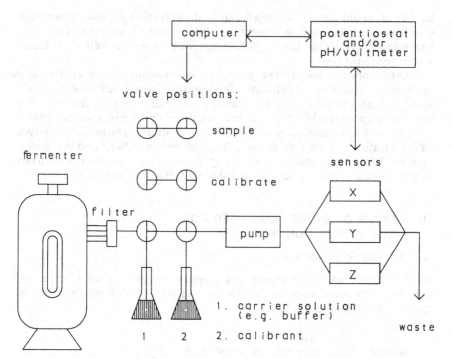

Figure 2   Detail of a flow-injection analysis system for monitoring fermentation.

macromolecules and microorganisms.   In addition, since most biosensors function best over a relatively narrow pH range, the in situ placement of the sensor may expose it to a nonoptimal or varying pH.

The two major problems encountered in adapting biosensors for the monitoring of bioprocesses are sterilizability and on-line calibration.   However, there has been some work aimed at circumventing these shortcomings.   It is possible to couple the sensor to the bioreactor (i.e., fermentation broth) so that the biosensor is now an external sensor.   The sample passes through a sterilizable membrane placed between the bioreactor and the biosensor.   Dialysis membranes have been shown to be useful for up to two weeks of operation in complex matrices [24], and have also been employed when dialyzable compounds such as carbohydrates, organic salts, and amino acids are present in the bioreaction [25].   Limitations of sterilizable membranes include membrane clogging, aging, and sampling lag time [2].   Also, the sterilization process for in situ sensors is typically done using steam at temperatures of 130°C for 0.5

h. This would result in irreversible deactivation of the sensor biocomponents. A possible alternative is the use of enzymes from
thermophilic bacteria [20]. These microbes survive well at boiling
water temperatures.

Other approaches to the sterility requirement for in situ sensors
have been examined. Penicillin fermentation was monitored in situ
with a sterile sensor with the enzyme penicillinase prepared by aseptic assembly methods [26]. Other enzyme- or microbe-based sensors have been chemically sterilized using dilute glutaric dialdehyde
[27], dilute diethylenetriamine [28], chloroform [26], and by autoclaving [29]. Microbiosensors for glucose have also been sterilized
using gamma radiation and enzymatic activity was retained [30].

III. TYPES OF BIOSENSORS USED FOR BIOPROCESS CONTROL

A. Enzymatic Biosensors

Most enzymes used for biosensors require oxygen as an electron acceptor. The most common is glucose oxidase (GOX) which oxidizes
glucose to gluconic acid:

$$\text{Glucose} + O_2 \xrightarrow{\text{GOX}} \text{gluconic acid} + H_2O_2$$

This reaction is monitored amperometrically either by measuring the
current generated by the increase of hydrogen peroxide or by the
decrease of oxygen. The limited solubility of oxygen in solution
can create a problem at high substrate concentrations often encountered in bioprocesses, such as fermentation, because it limits the
linear range of the sensor at higher concentrations.

Figure 3 shows a schematic of the enzymatic reaction at the
electrode surface, oxygen regenerates the oxidized, active form of
the enzyme. Oxygen is regenerated from hydrogen peroxide at the
electrode surface which is at a positive potential, however, this regenerated oxygen is not enough to meet the demands of the enzymatic reaction at high substrate concentrations due to losses from
diffusion.

Electron transfer mediators, such as benzoquinone or ferrocene,
can take the place of oxygen. Currently, much effort is directed
at developing a sensor that uses an electron mediator in place of
oxygen, thus overcoming the problem of limited availability of oxygen. Compounds that have been commonly used as mediators are
benzoquinone [31] and ferrocene derivatives [32–34]. It has been
shown that the use of benzoquinone contributes less to the deacti-

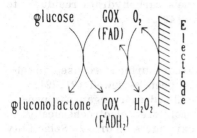

Figure 3   Schematic of the enzymatic reaction taking place at the electrode surface of the biosensor.

vation of the enzyme than the reduction of oxygen to hydrogen peroxide, thereby increasing the operating lifetime of an enzyme biosensor [35]. Also the focus of recent research is the coupling of an electron mediator with a species that electrochemically polymerizes to form a film over the electrode that both mediates electron transfer, and serves to screen out interferents and prevent electrode fouling [36].

Another area of interest is the use of films or membranes to protect the biosensor surface [37,38]. These are especially useful for amperometric electrodes because a film could screen out unwanted electroactive compounds that would otherwise interfere with the measurement. Materials for films include cellulose acetate membranes, and electropolymerized compounds such as 1,2-diaminobenzene and its isomers [37], and salicylate derivatives [39]. Films or membranes also help potentiometric and amperometric biosensors by screening out biomacromolecules (e.g., proteins) that are not electrochemical interferents but fouling agents due to their tendency to adsorb onto electrode surface. This causes a variety of problems because the fouled surface no longer behaves well electrochemically since fouling changes the surface area of the electrode and hinders rapid diffusion of analytes to the sensing surface.

Ideally, the immobilized enzyme on the sensor would not be denatured by being subjected to elevated temperatures required for sterilization. Therefore, research into methods of further stabilizing immobilized enzymes [40–43], or synthesizing thermally stable artificial enzymes [44–48] is an important area of research and has received considerable attention recently. For example, Barbaric et al. were able to improve the stability of glucose oxidase through oxidation, with periodate, of the enzyme's carbohydrate residuals to carbonyl groups prior to immobilization [49].

B. Microbial Sensors

Microorganisms immobilized on the electrode surface are used in microbial biosensors. These sensors have been reviewed [21,22]. The respirational activities of the immobilized microorganisms produce certain metabolites which can be monitored electrochemically. These include hydrogen gas or ions, oxygen, and urea. Selectivity is obtained by choice of the microorganism. For example, oxygen is consumed by an organism as it assimilates a variety of organic compounds. By monitoring with an oxygen electrode, one sees a current decrease corresponding to a decrease in local oxygen concentration around the electrode surface. The consumption of oxygen by the microorganism and the diffusion of oxygen to the membrane will reach a steady state indicated by a constant current level. Microbial biosensors are used for the determination of substrates and activities of enzymes, for estimating the biological oxygen demand, and for characterizing the physiological state of the microbe during fermentation [50]. Vais and Margineanu have developed a kinetic model of electrochemical biosensors using immobilized bacteria [51].

C. Tissue Sensors

Tissue-based biosensors include a wide variety of both animal and plant tissues as the biochemical transducer. Fewer than 10 years ago, only animal tissues, mostly from mammals, were used in the construction of biosensors. The first report of a tissue membrane electrode used beef liver tissue coupled with urease on an ammonia-sensitive electrode for the monitoring of arginine [52]. The conversion of arginine to ornithine by the enzymes in the tissue produces urea. Urease converts urea to ammonia and carbon dioxide. The addition of azide to any buffer solution prevents deterioration of the tissue and provides for lifetimes of over a month. Plant tissures were not believed to contain a sufficient level of activity needed for a sensor. Rechnitz has used young leaves, root tips, blossoms, and seeds in his research and has shown that these plant parts will function in a biosensor [53].

Construction of these biosensors involves the coupling of a specific tissue with a gas-sensing electrode. The reaction of interest is then monitored potentiometrically. According to Rechnitz, future efforts will focus on "devising sensors which can selectively measure several substrates using the same tissue under different operating conditions" [53].

IV. APPLICATIONS

The vast majority of work reported using electrochemical biosensors to monitor bioprocesses involves fermentation. Fermentation is a general term for the degradation of organic compounds into simpler products to generate adenosine triphosphate (ATP). Electrochemical biosensors already exist that can detect the compounds involved in fermentation very successfully. Also, FIA has successfully been applied to biosensor research (see any of the previously mentioned reviews).

Kok and Hogan have designed and tested probe calibrators to allow for in situ calibration of sensors in fermentation broth [54]. Their configuration allows an oxygen electrode to be calibrated and cleaned in place by isolating from the broth. Once the sensor is ready, it is again exposed to the fermentation broth. Also, if desired, the sensor can be replaced by a new electrode, which is then cleaned, calibrated, and used. The use of these probes allows the changing calibration constants of the biosensors, as a result of fouling, to be taken into account. They cite two important benefits of using a calibrator probe. First, the sensor is cleaned and calibrated in situ; second, a new sensor can be installed, sterilized, and calibrated without contamination of the broth. The sensor used was an oxygen electrode but the researchers claim the techniques can be applied to other sensors.

A. Fermentation

Electrochemical biosensors hold much promise for monitoring many of the important compounds involved in fermentations. These include glucose, ethanol and methanol, organic acids, and various gases. A general review of fermentation was done by Davis et al. [55]. Some work has been directed at utilizing biochemical parameters that can be determined in fermentations by already established equations. Merton et al. have studied methods for on-line determination of biochemical and physiological parameters either in continuous or discontinuous modes during fermentation, as well as covering types of biosensors, calibration of sensor systems, and applications [56]. A theoretical treatment of enzyme stability during continuous use in bioreactors has been given [57].

Monitoring Glucose

Glucose is monitored amperometrically using the enzyme glucose oxidase as the biochemical transducer. Glucose dehydrogenase could be used but the addition of NAD as a cofactor is required. Glucose oxidase is routinely immobilized on a platinum or carbon surface by crosslinking with glutaraldehyde or covalent bonding to functional groups present on the electrode surface [4].

Karube et al. reported the successful use of a microbial sensor for glucose using whole bacteria cells [58]. Linear concentration ranges were obtained for glucose solutions of 2—20 mg/L and the sensor was good for two weeks and 150 assays. The sensor was also successfully used for the analysis of glucose in molasses. Although no bioprocess application was done at the time, the authors cited the potential for use in microbial cultures and many reports in this area follow Karube's lead.

Recently, Mandenius used glucose oxidase and catalase in a packed-bed reactor coupled to an immersible dialysis probe for continuous amperometric monitoring of glucose in a fermentation broth [59]. This was in situ monitoring where glucose was sampled through a dialysis probe immersed in the broth. Six different membranes were used in the probe, and the extent of yeast cell adhesion, as well as the correlation between glucose transfer and membrane area were studied for each. Membranes included regenerated cellulose, cuproferious regenerated cellulose (Cuprofane), polypropene, cellofane, and polyamide. A Clark oxygen electrode was used to monitor oxygen depletion. Linear calibration curves of 1—60 m$\underline{M}$ glucose were obtained.

Using immobilized glucose oxidase on an oxygen electrode and monitoring oxygen depletion, Su and Chen constructed a glucose sensor for fermentation monitoring [60]. The paper describes the fabrication and the characterization of the sensor when used in batch mode, i.e., electrochemical analyses were done off-line. They achieved a sampling rate of 20/h.

Research into simultaneous determination of glucose and sucrose using enzyme electrodes has also been done by Geppert and Asperger [61]. Work was aimed at constructing a fully automated online method for measurement of different fermentation substrates. Rapid response, large concentration ranges, and no prior separation of biomass were also goals of the research. A discussion of the characterization of enzyme electrodes is given using basic transport equations and electrode kinetics. Conclusions drawn from this characterization are that there is a specific electrode geometry which results in optimal transport properties, the immobilized enzyme layer should contain a high level of enzyme to offset any decrease in activity that may occur over time, and the electrode should have low resistance. The analysis was completely automated, but manual operation was used if necessary for optimization.

Geppert and Asperger substituted benzoquinone for oxygen as the electron acceptor in the enzymatic reaction studied. A marked increase in linear range by measuring the reduction of benzoquinone versus monitoring either hydrogen peroxide oxidation or oxygen consumption was shown using three electrodes, one utilizing each of the three monitoring modes. The inferiority of the two methods that depend on the ambient concentration of oxygen is due to a lack of oxygen needed for the enzymatic reaction at higher glucose concentrations. Also reported in this work is the successful simultaneous determination of glucose and sucrose. Two electrodes were used, one with an enzyme for glucose and the other with enzymes for both glucose and sucrose. The glucose signal from the first electrode is electronically subtracted from the additive signal from the second to give a signal for sucrose [61].

Clarke et al. modifying existing enzyme sensor technology, overcame three problems common to in situ monitoring over time: insufficient stability, increasing response times, and decreasing linear concentration range [62]. Two novel methods were incorporated in the sensing system designed to overcome these problems. First, glucose oxidase, prior to immobilization, was treated with sodium periodate, oxidizing the enzyme's carbohydrate residues to carbonyls. Second, increased lifetime was gained by coating the electrode with alkylamine. These two modifications led to a sixfold increase in initial response over the same sensor without the two modifications. This novel sensor was used successfully to monitor glucose levels in a batch culture of E. coli.

Mizutani et al. have developed an automatic glucose analyzer for on-line control of glucose concentration [63]. The biosensor consisted of glucose oxidase immobilized on an oxygen electrode. This method allowed choosing a glucose concentration desired for the fermentation broth, analyzing the broth for glucose on-line at chosen intervals and, using a fed-batch system, adjusting the broth to the desired glucose level. When 0.3 g/L was chosen, the system kept the glucose level in the broth between 0.08 and 0.54 g/L and for a set concentration of 10 g/L, 9–11 g/L was observed. Proposed means of overcoming difficulties at lower glucose or higher cell concentrations are discussed. Also pointed out is the need to keep sampling intervals as short as possible.

Monitoring Ethanol and Methanol

Methods of ethanol detection and future directions for the development of enzyme biosensors in general have been reviewed by Wiseman [64]. A review on synthetic and analytical applications of immobilized alcohol dehydrogenase has also been published by Wiseman [65]. Hopkins has described the construction of a multipurpose enzyme sensor based on alcohol oxidase [66].

Karube et al. developed separate microbial sensors for the online determination of ethanol and methanol in fermentation broths

[67]. Selectivity was achieved using bacteria that respond to one
but not the other alcohol. For ethanol, Trichosporon brassicae was
used. The microorganism specific for methanol was not identified
by the authors. An oxygen electrode covered with a gas-permeable
Teflon membrane, nylon net, and a porous acetyl cellulose membrane
containing the adsorbed microorganisms were used. Neither the
sensor constructed for detection of ethanol nor methanol responded
to other compounds present in the broth such as formic acid, acetic
acid, carbohydrates, amino acids, and ions. The sensors were useful for more than three weeks and 2100 assays with acceptable linear
calibration curves (r = 0.98) between 2.0 and 22.5 mg/L. The results obtained using the microbial sensors compared well with those
obtained using gas chromatography.

     The lack of a commercially available alcohol oxidase which is
both stable and highly specific has led researchers to attempt to obtain the enzyme in pure form from a strain of yeast [68]. Oxygen
depletion from the oxidation of methanol or ethanol to peroxide and
formaldehyde or acetaldehyde was monitored with an O_2 sensor on
which a gelatin matrix containing the immobilized enzyme is coated.
A very good response time of <2 min was reported, and the sensor
was stable over 500 assays. The range for methanol was 0.5 to 15
mM, and for ethanol 10 to 300 mM. Beer and wine were analyzed
for ethanol content, and the authors claim that the use of their enzyme sensor gives preferable results to chemical methods of analysis
(spectroscopic) since color does not interfere. However, it should
be pointed out that alcohol oxidase responds to both ethanol and
methanol, and one could not be accurately monitored if the other ex-
isted to any appreciable extent. The yeast production of ethanol
from glucose was monitored with the same sensor in a closed-loop,
automated system.

     A semiconductor gas-sensing electrode with immobilized cells has
been used to determine ethanol levels in fermentation broth [69].
The sensor was designed so that monitoring was done with the sensor dipped directly in the broth. The same paper reported a second
microbial biosensor used to monitor indole during the fermentation of
L-tryptophan. This sensor is used instead in a fed-batch process
with E. coli as the immobilized bacteria.

Monitoring Organic Acids

     The monitoring of acetic acid in bioreactors is important because
acetic acid is often the carbon source for microorganism cultivation.
However, if the concentration of acetic acid is too high, it inhibits
growth. Thus the ability to maintain the optimal acetic acid concentration through on-line monitoring is desirable.

     Using the same biosensing system used for ethanol monitoring,
Karube et al. monitored acetic acid levels in a fermentation broth of

glutamic acid [70]. It was necessary to keep the pH of the solution below 4.75 (pK$_a$ of acetic acid), because acetate ions do not pass through the gas-permeable membrane. Oxygen depletion was monitored as acetic acid was assimilated by the bacteria. The sensor had a response time of 8 min, a linear range of 2—54 mg/L, and performed well for three weeks and 1500 assays.

Glutamic acid can be monitored using glutamate decarboxylase or a microorganism that contains this enzyme. E. coli has been used for this purpose by immobilization on a CO_2-sensing electrode. Decarboxylation of glutamic acid produces CO_2 [71]. The sensor exhibited a linear range of 100—800 mg/L with response time of under 5 min, and a lifetime of over three weeks with 1500 assays.

Monitoring Gases

There have been several reports on the use of biosensors for monitoring ammonia gas in processes which do not use ammonia gas electrodes [72—74]. These efforts were aimed at circumventing interferences in the use of ammonia gas sensors. Instead, certain bacteria that use ammonia an energy source, consuming oxygen in the process, were immobilized on an O_2 electrode. The use of these "nitrifying" bacteria with an oxygen electrode to monitor ammonia operates the same way as the previous applications of oxygen electrodes. Ammonia levels of 0.1 to 42 mg/L produced linear calibration curves with a 4-min response time, a lifetime of 10 days and 200 assays. The sensor showed very good selectivity to ammonia.

Methane has been determined using an oxygen electrode coupled with immobilized M. flagellata which uses methane as its source of energy. The utilization of methane by the microbe consumes oxygen [75,76]. The response time of the sensor system was 1 min, and a linear range of 13.1 μM to 6.6 mM was observed. The results obtained with this methane microbial sensor compared well with those determined by GC.

Miscellaneous

Turner has reported the construction and successful use of amperometric enzyme electrodes for several fermentation analytes [77]. The sensor constructed for monitoring glucose using glucose oxidase was by far the best, having a half-life of 600 h. Others tested had half-lives of 1.5—24 h. Carbon paste or Pt foil were used as electrodes, and the reduction of a ferrocene mediator was proportional to glucose concentration was measured.

Gibson and Woodward used spectroscopic means as well as a commercial peroxide sensor to monitor both glucose and ethanol in fermentation broth [78]. Using Saccharomyces cerevisae, and attempting to mimic standard brewing condtions, they suggested such a system would find use in the beverage industry.

Electrochemical biosensors also hold much potential for the determination of antibiotics produced by bacteria in fermentation processes. A review of the analysis and control of antibiotic production has been given recently by Schügerl [79]. Karube has reported a method for the monitoring of cephalosporin by immobilizing the microbe Cirobacter freundii on a pH glass electrode [21]. This bacterium produces cephalosporinase which acts on cephalosporin and generates H^+ ions. The potential versus log of cephalosporin concentration gave a linear curve, which enabled determination of 7-phenylacetylamidodesacetoxysporanic acid, cephaloridine, cephalothin, and cephalosporin c. Response time was about 10 min and showed no decrease in performance after a week with several runs each day. Results correlated well with results obtained by high performance liquid chromatography (HPLC).

Controlled-pore glass embedded in plastic tubing as the basis for an enzyme reactor has been used for the determination of penicillin using penicillinase in a continuous flow stream [80]. The pH changes are monitored with a glass pH electrode modified with immobilized penicillinase. These reactors are constructed quickly and easily, exhibit very good uniformity, and high local activity. Tygon and Teflon tubing were used and evaluated as materials containing the flow-through controlled-pore glass. The results reported show that this configuration gives an increased response to penicillin compared to an etched borosilicate glass open tubular reactor (OTR). The authors compare the responses to 0.25, 0.50, and 1.00 mM penicillin for the OTR, and for porous glass reactors with pore diameters of 0.75 and 30 μm. The larger pore reactor showed sensitivity three times that of the OTR for the above range. This work holds promise for the monitoring of penicillin or other species of interest in bioprocesses. The fermentation of malic acid to form lactate in wine has been controlled by monitoring the decarboxylation of L-malic acid to L-lactic acid by immobilizing various heterolactic bacteria [81].

B. Electrochemical Biosensors for Other Processes

A method for the simultaneous determination of sucrose, glucose, and fructose would find much use in bioprocesses. Matsumoto et al. have developed a FIA system using a parallel configuration of enzyme reactors [82]. Both glucose and sucrose are determined amperometrically using glucose oxidase, where oxidation of hydrogen peroxide is measured. The sucrose reactor includes a glucose-eliminating reactor coupled to the reactor for sucrose so that only sucrose enters the sucrose reactor. Glucose entering the sucrose reactor would be an interferent. Ascorbic acid is an electroactive interferent at positive working potentials, therefore an ascorbate eliminating reactor with immobilized ascorbate oxidase removed vitamin C. Fructose was determined with immobilized fructose 5-dehydrogenase using hexacyanoferrate(III) as the redox mediator. The sensor system gave linear responses of 0.02–1.0 m$\underline{M}$ (r > 0.999) for each sugar.

Monitoring of lactose in processes is important and the use of sensors for lactose in meat products has been discussed [83]. A two-enzyme biosensor has been used to determine lactose in raw milk [84]. The same workers also report using a three-enzyme system (β-galactosidase, lactate oxidase, and glucose oxidase) together with a peroxide sensor for the in-line determination of lactose, lactate, and glucose during milk processing [85]. By coimmobilizing lactate dehydrogenase and glutamic-pyruvic transaminase on a porous silica support, Gorton and Hedlund developed an FIA system for determination of L-lactate. Meldola blue, an organic dye, replaces oxygen as the electron acceptor, allowing a working potential of 0 V (versus Ag/AgCl). This is advantageous since most electrochemical interferents are not electroactive at this potential [86].

V. CONCLUSIONS AND FUTURE DIRECTIONS

Up to this point, the application of electrochemical biosensors to bioprocesses has only started. Future research efforts will be of two types: those directed at improving in situ methods of monitoring and those aimed at optimizing FIA and closed-loop systems. There are common and unique problems to each. Examples were given in the introduction to this chapter. These will, no doubt, be dealt with in the near future. For a closed-loop monitoring system where the sensor is external to the reaction broth, research into "cleaning up" the broth sample before it reaches the biosensor is needed. This involves development of new and improved membranes for external monitoring systems. For in situ sensors, the most pressing issue is protecting the biochemical/electrochemical couple from being damaged by species in the reaction broth. Avoiding adsorption of biomacromolecules onto the sensing layer is of utmost importance, as fouling of this surface results in poor electrochemical behavior of the electrode. Electropolymerized films hold much promise because they protect the electrode and are generally very thin, thus minimizing the diffusion time of analytes from the broth to the sensor.

The area of synthetic enzymes will be important, especially the synthesis of enzymes that are stable at the temperatures needed for steam sterilization. The gains made from this area of biochemistry will no doubt hold much promise for the area of enzyme biosensors.

Response time and working lifetime of the biosensor are two parameters that can always be improved upon. No doubt much effort will be directed toward improving both of these. Another important area that can be improved is calibration of the sensor. Calibration needs to be done quickly and as often as required. Ideally, calibration should be done automatically. In closed-loop external bioreactors this would involve a switching system that periodically stops flow of the broth through the bioreactor and allows flow of solutions necessary for calibration. Calibration being complete, the system would switch back to monitoring the broth. For in situ sensors, whose lifetimes are limited anyway, the use of precalibrated disposable sensors seems a good route, but currently a prohibitively expensive one.

Despite these difficulties, it is clear that the field of bioprocess control will use biosensors to a much greater degree in the future. Biosensors present an attractive means for improving productivity and monitoring product quality.

REFERENCES

1. Duffy, P., Saad, I., and Wallach, J. M., <u>Anal. Chim. Acta</u>, <u>213</u>(1–2), 267 (1988).
2. Twork, J. and Yacynych, A. M., <u>Biotechnol. Prog. 1986</u>, <u>2</u>(2), 67 (1986).
3. Karube, I., <u>Biotechnol. Genet. Eng. Rev.</u>, <u>2</u>, 313 (1984).
4. Weibel, M. K., Dritschilo, W., Bright, H. J., and Humphrey, A. E., <u>Analyt. Biochem.</u>, <u>52</u>, 402 (1973).
5. Nakamura, K., Nankai, S., and Iijima, T., Patent No. 4392933, Matushita Electric Industrial Co. Ltd. (1983).
6. Yao, T. and Musha, S., <u>Anal. Chim. Acta</u>, <u>110</u>, 203 (1979).
7. Rechnitz, G. A., <u>TrAC</u>, <u>5</u>(7), 172 (1986).
8. Rechnitz, G. A., <u>Chemical and Engineering News</u>, <u>66</u>(36), 24 (1988).
9. Frew, J. E. and Hill, H. A. O., <u>Anal. Chem.</u>, <u>59</u>, 933A (1987).
10. Kobos, R. K., <u>TrAC</u>, <u>6</u>(1), 6 (1987).
11. Koryta, J., <u>Electrochimica Acta</u>, <u>31</u>, 515 (1986).
12. Hall, E. A. H., <u>Int. J. Biochem.</u>, <u>20</u>, 357 (1988).
13. Romette, J. L. and Thomas, D., <u>Methods Enzymol.</u>, <u>137</u>, 44 (1988).
14. Guilbault, G. G. and Kauffmann, J., <u>Biotechnol. Appl. Biochem.</u>, <u>9</u>(2), 95 (1987).
15. Foulds, N. C. and Lowe, C. R., <u>BioEssays</u>, <u>3</u>(3), 129 (1985).
16. Clarke, D. J., Calder, M. R., Carr, R. J. G., Blake-Coleman, B. C., Moody, S., and Collinge, T. A., <u>Biosensors</u>, <u>1</u>(3), 213 (1985).

17. Clarke, D. J., Phil. Trans. R. Soc. Lond. B., 316, 169 (1987).
18. Clarke, D. J. and Sherwood, R. F., Chim. Oggi., (3), 51 (1987).
19. Schultz, J. S. and Meyerhoff, M., Enzyme Microb. Technol., 9, 697 (1987).
20. Orr, T., Genet. Engin. News, 8(9), 27, 40-1 (1988).
21. Karube, I., in Fundamentals and Applications of Chemical Sensors (D. Schuetzle and R. Hammerle, eds.), ACS Symposium Series (1986).
22. Karube, I., Tamiya, E., Sode, K., Yodoyama, K., Kitagawa, Y., Suzuki, H., and Asano, Y., Anal. Chim. Acta, 213(1–2), 69 (1988).
23. Enfors, S. O. and Cleland, N., Methods Enzymol., 137, 298 (1988).
24. Mandenius, C. F., Danielsson, B., and Mattiasson, B., Anal. Chim. Acta, 163, 135 (1984).
25. Zabriskie, D. W. and Humphrey, A. E., Biotech. Bioengin., 20, 1295 (1978).
26. Hewetson, J. W., Jong, T. H., and Gray, P. P., Biotech. Bioengin. Symp. No. 9 (1979).
27. Weetall, H. H., Havenala, N. B., Pitcher, W. H., Detar, C. C., Venn, W. P., and Yaverbaum, S., Biotech. Bioengin., 16, 689 (1974).
28. Baret, J. L., Patent No. 4,393,138, Corning Glass Works (1983).
29. Enfors, S. O. and Nilsson, H., Enzyme Microb. Technol., 1, 260 (1979).
30. Churchouse, S. J., Battersby, C. M., Mullen, W. H., and Vadgama, P. M., Biosensors, 2, 325 (1986).
31. Cenas, N. K., Pocius, A. K., and Kulys, J., Bioelectrochem. Bioenerg., 11(1), 61 (1983).
32. Cass, A. E. G., Davis, G., Francis, G. D., Hill, H. A. O., Aston, W. J., Higgins, I. J., Plotkin, E. V., Scott, L. D. L., and Turner, A. P. F., Anal. Chem., 56(4), 667 (1984).
33. Foulds, N. C. and Lowe, C. R., Anal. Chem., 60, 2473 (1988).
34. Dicks, J. M., Aston, W. J., Davis, G., and Turner, A. P. F., Anal. Chim. Acta, 182, 103 (1986).
35. Bourdillon, C., Hervgault, C., and Thomas, D., Biotech. Bioeng., 27, 1619 (1985).
36. Frew, J. E., Harmer, M. A., Hill, H. A. O., and Libor, S. I., J. Electroanal. Chem., 201(1), 1 (1986).
37. Geise, R. J. and Yacynych, A. M., 1988 ACS National Meeting (in press).
38. Merz, A. and Bard, A. J., J. Am. Chem. Soc., 100(10), 3222 (1978).

39. Mourcel, P., Pham, M., Lacaze, P. C., and Dubois, J., J. Electroanal. Chem., 145, 467 (1983).
40. Klibanov, A. M. and Mozhaev, V. V., Biochem. Biophys. Res. Commun., 83, 1012 (1978).
41. Klibanov, A. M., Science, 219(4585), 722 (1983).
42. Klibanov, A. M., Adv. Appl. Microbiol., 29, 1 (1983).
43. Lenders, J. P. and Crichton, R. R., Biotechnol. Bioengin., 26, 1343 (1984).
44. Kaiser, E. T., Pure Appl. Chem., 56(8), 979 (1984).
45. Kaiser, E. T. and Lawrence, D. S., Science, 226(4674), 505 (1984).
46. Zimmerman, S. C. and Breslow, R., J. Am. Chem. Soc., 106(5), 1490-1 (1984).
47. Czarnik, A. W. and Breslow, R., Carbohydr. Res., 128(1), 133 (1984).
48. Wulff, G., Best, W., and Akelah, A., React. Polym. Ion. Exch., Sorbents, 2(3), 167 (1984).
49. Barbaric, S., Kozulic, B., Leustek, I., Pavlovic, B., Cesi, V., and Mildner, P., 3rd Eur. Congr. Biotechnol., 1, 307 (1984).
50. Riedel, K., Renneberg, R., Liebs, P., and Kaiser, G., Stud. Biophys., 119(1—3), 163 (1987).
51. Vais, H. and Margineanu, D. G., Bioelectrochem. Bioenerg., 16(1), 5 (1986).
52. Chem. Eng. News, 56(41), 16 (1978).
53. Sidwell, J. S. and Rechnitz, G. A., Biosensors, 4(2), 221 (1986).
54. Kok, R. and Hogan, P., Biosensors, 3(2), 89 (1987).
55. Davis, C. R., Wibowo, D., Eschenbruch, R., Lee, T. H., and Fleet, G. H., Am. J. Enol. Vitic., 36(4), 290 (1985).
56. Merton, O. W., Palfi, G. E., and Steiner, J., J. Adv. Biotechnol. Processes, 6, 111 (1986).
57. Yamane, T., Siriote, P., and Shimuzu, S., Biotechnol. Bioeng., 30(8), 963 (1987).
58. Karube, I., Mitsuda, S., and Suzuki, S., Euro. J. Appl. Microbiol. Biotechnol., 7, 343 (1979).
59. Mandenius, C. F., Anal. Lett., 21(10), 1817 (1988).
60. Su, Y. C. and Chen, C. Y., Proc. Natl. Soc. Counc. Repub. China, Part B: Life Sci., 11(1), 10 (1987).
61. Geppert, G. and Asperger, L., Bioelectrochem. Bioenerg., 17(3), 399 (1987).
62. Brooks, S. L., Ashby, R. E., Turner, A. P. F., Calder, M. R., and Clarke, D. J., Biosensors, 3(1), 45 (1987).
63. Mizutani, S., Iijima, S., Ogawa, Y., Izumi, R., Matsumoto, K., and Kobayashi, T., J. Ferment. Technol., 65(3), 325 (1987).

64. Wiseman, A., TrAC, 7(1), 5 (1988).
65. Wiseman, A., in Topics in Enzyme and Fermentation Biotechnology, vol. 5 (A. Wiseman, ed.), Ellis Horwood, Chichester (1981), pp. 337–354.
66. Hopkins, T. R., Am. Biotechnol. Lab., 3(5), 32 (1985).
67. Hikuma, M., Kubo, T., Yasuda, T., Karube, I., and Suzuki, S., Biotechnol. Bioengin., 21, 1845 (1979).
68. Belghith, H., Romette, J. L., and Thomas, D., Biotechnol. Bioeng., 30(9), 1001 (1987).
69. Vorlop, K. D., Becke, J. W., Stock, J., and Klein, J., 3rd Eur. Congr. Biotechnol., 2, 325 (1984).
70. Hikuma, M., Kubo, T., Yasuda, T., Karube, I., and Suzuki, S., Anal. Chim. Acta, 109, 33 (1979).
71. Hikuma, M., Obana, H., Yasuda, T., Karube, I., and Suzuki, S., Anal. Chim. Acta, 116, 61 (1980).
72. Karube, I., Okada, T., and Suzuki, S., Anal. Chem., 53(12), 1852 (1981).
73. Okada, T., Karube, I., and Suzuki, S., Anal. Chim. Acta, 135, 159 (1982).
74. Hikuma, M., Kubo, T., Yasuda, T., Karube, I., and Suzuki, S., Anal. Chem., 52(7), 1020 (1980).
75. Okada, T., Karube, I., and Suzuki, S., Europ. J. Appl. Microbiol. Biotechnol., 12(2), 102 (1981).
76. Karube, I., Okada, T., and Suzuki, S., Anal. Chim. Acta, 135, 61 (1982).
77. Turner, A. P. F., World Biotech Rep., 1, 181 (1985).
78. Gibson, T. D. and Woodward, J. R., Anal. Chim. Acta, 213(1–2), 61 (1988).
79. Schügerl, K., Anal. Chim. Acta, 213(1–2), 1 (1988).
80. Gosnell, M. C., Snelling, R. E., and Mottola, H. A., Anal. Chem., 58(7), 1585 (1986).
81. Crapisi, A., Spettoli, P., Nuti, M. P., and Zamorani, A., Appl. Bacteriol., 63(6), 513 (1987).
82. Matsumoto, K., Kamikado, H., Matsubara, H., and Osajima, Y., Anal. Chem., 60(2), 147 (1988).
83. Ellis, P. C. and Rand, A. G., J. Assoc. Anal. Chem., 70(6), 1063 (1987).
84. Pilloton, R., Mascini, M., Casella, I. G., Festa, M. R., and Bottari, E., Anal. Lett., 20(11), 1803 (1987).
85. Mascini, M., Moscone, D., Palleschi, G., and Pilloton, R., Anal. Chim. Acta, 213(1–2), 101 (1988).
86. Gorton, L. and Hedlund, A., Anal. Chim. Acta, 213(1–2), 91 (1988).

9

Thermistor Probes

NEIL D. JESPERSEN   *St. John's University, Jamaica, New York*

I. INTRODUCTION

Thermistors are recognized as very sensitive temperature sensors.
As such, they have found a great deal of use in a wide variety of
chemical applications. Although their use is well known as thermal
conductivity detectors in gas chromatography or as infrared radiation detectors in spectroscopy, they are also widely used in solution
calorimetry and thermometric titrimetry. In all of these applications,
thermistors are recognized as universal detectors with no inherent
chemical selectivity of their own. Any process or chemical reaction
that produces or absorbs heat can be detected. It is up to the in-
vestigator to properly control physical processes and ingeniously
manipulate chemical characteristics of a system under study in order
to obtain useful results. Thermal probes are examples of this art-
ful manipulation. Several reviews of thermistor probes are available
[1-7].

Solution calorimetry and thermometric titrimetry are beyond the
scope of this work. However, the basic principles of thermistor
probes stem from much of the work done in these fields. Therefore, the first section is devoted to basics of thermistor operation
and the limitations of thermistor measurements.

Some thermistor probes were designed to mimic potentiometric
probes such as the glass electrode. Direct probes of this sort are
described in the next section of this chapter. These were generally
successful in showing that useful data could be obtained from a fairly simple and straightforward manipulation of the chemical and physical environment.

Another set of thermistor probes were developed around flow
systems. In this case, the thermistor monitored a specific reaction
taking place in a column just before the thermistor. The obvious
ease and rapidity of analysis was of interest here along with the ap-
plicability to continuous process measurements.

In the last section of this chapter the reported applications of these devices are summarized. Each of the many thermistor probe devices has distinct characteristics which make it suitable for use in specific cases which are outlined here.

II. THERMISTOR CONSTRUCTION, THEORY AND OPERATION

A. Construction

Thermistors are constructed from mixtures of metallic oxides such as cobalt, copper, manganese, nickel, iron, and uranium which are sintered at high temperatures into ceramics. Electrical characteristics are controlled by the specific chemical mixtures used and the physical size and configuration of the finished product. Finished thermistors may be in the form of small beads, approximately 0.2 to 1.2 mm in diameter, or they may be formed in the shape of rods, disks or washers. These may then be encapsulated in a protective material or left uncoated. Typically, the bead-encapsulated-in-glass configuration is used in thermistor probe studies. Although encapsulation acts as insulation and tends to decrease the speed of thermistor response, it also is inert to the chemical environment in which they are used.

To produce a bead, two fine wires are held parallel approximately 0.02 mm apart, and a small ellipsoid of thermistor material is formed around the wires. Upon sintering, the wires become imbedded in the bead with good electrical contact. These beads are then made into probes by sealing them into glass rods.

B. Temperature Response and Measurement

Response of thermistors to temperature change follows classical semiconductor theory. In the solid, the conduction band and the valence band are separated by a relatively small energy gap. This energy gap is small enough that changes in temperature will significantly change the number of electrons energetic enough to enter the conduction band. An increase in temperature increases the number of electrons in the conduction band and therefore decreases the resistance of the thermistor bead. This is adequately described by the equation [8]

$$\frac{R_1}{R_2} = \exp B\left(\frac{1}{T_1} - \frac{1}{T_2}\right) \tag{9.1}$$

where R_1 and R_2 represent the resistance of the thermistor at temperatures T_1 and T_2, respectively, and B is a constant typically having a value of 3000 to 5000. This expression leads to an exponentially decreasing resistance for the thermistor as temperature increases.

Another figure of interest for the thermistor is the percent resistance change per degree Celsius, a_t. This is usually on the order of -4% °C^{-1} for a typical thermistor at room temperature. It compares very favorably to a platinum resistance thermometer where the value is 0.36% °C^{-1} [8].

Since electrical resistance is the property which varies with temperature, thermistors are commonly incorporated into Wheatstone bridge circuits for measurement purposes. Precise measurements of resistance may be obtained if the Wheatstone bridge shown in Figure 1 is balanced manually so that no electrical potential exists between points A and B. In this balanced condition, it is known that

$$\frac{R_1}{R_T} = \frac{R_2}{R_3} \tag{9.2}$$

R_T represents the resistance of the thermistor, and the other resistances are part of the Wheatstone bridge. For the best sensitivity, all resistances should be approximately equal.

In actual practice, it is very inconvenient, if not impossible, to balance the Wheatstone bridge, and therefore the unbalance potential is monitored by a strip chart recorder or on-line computer. The expression relating the unbalance potential to the thermistor resistance is

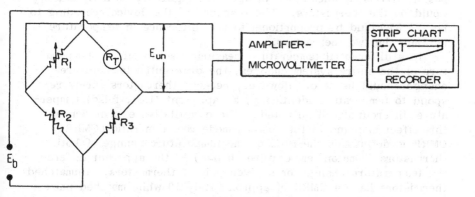

Figure 1   A typical Wheatstone bridge apparatus. E_b is the applied potential to the bridge, and E_{un} is the unbalance potential. R_T represents a thermistor in a single thermistor device. R_1 and R_T are thermistors in a differential apparatus.

$$E_{un} = E_b \left(\frac{R_1}{R_1 + R_2} - \frac{R_T}{R_T + R_3} \right) \qquad (9.3)$$

where E_{un} and E_b represent the unbalance potential and the input bridge potential, respectively.

It is interesting to note that the nonlinearity of the thermistor response with temperature, Eq. (9.1) is cancelled somewhat by the nonlinearity of the Wheatstone bridge unbalance potential, Eq. (9.3). Using typical thermistor values for Eq. (9.1) and substituting this into Eq. (9.3), this compensation effect may be readily demonstrated. Using a 100,000-Ω thermistor and the appropriate constants, linear regression analysis of plots of thermistor resistance, bridge potential (assuming linear resistance change), and bridge potential using the actual thermistor resistance, all as a function of temperature indicate that the standard error of estimate is 0.0266 for the thermistor-bridge combination. For the thermistor resistance alone or the bridge performance with a linear resistance change, the standard errors of estimate are 0.180 and 0.152, respectively. Correlation coefficients also bear out this observation with the three values being 0.9999996, 0.99998, and 0.999986, respectively. Clearly, the nonlinearity of the Wheatstone bridge compensates for the nonlinearity of the thermistor resistance change.

C. Differential Temperature Measurement

In many instances it is desirable to make differential temperature measurements. This can be done if two thermistors are used in opposing arms of a Wheatstone bridge. In Figure 1, both R_1 and R_T would be the thermistors. The response of the device according to Eq. (9.3) would be proportional to the difference in temperature between the two thermistors.

In a differential temperature measuring setup, if the temperature of the entire system changes, the differential temperature change should be zero. However, because thermistors rarely respond to temperature identically, an apparent (i.e., false) temperature differential will be noted. The quantitative explanation of this effect is found in the common-mode rejection ratio (CMRR). CMRR is defined as the ratio of the temperature change of both thermistors (common temperature change) to the apparent differential temperature change for a given pair of thermistors. Unmatched thermistors have a CMRR of approximately 10 while matched pairs can have CMRR values over 1000. Temperature changes which affect both thermistors should result in no apparent ΔT if the CMRR is infinite. A CMRR of 10 indicates that an apparent ΔT of 0.1°C will be observed if both thermistors are heated by 1°. Similarly, a CMRR of 1000 will result in a ΔT of one millidegree for every one degree rise in common temperature. Since one of the major objectives of differential temperature measurements is to decrease thermostating requirements, well matched thermistors should be used or the effort will be self-defeating [9].

D. Thermistor Time Constant

In addition to temperature response, the time response of a thermistor is of interest in the development of thermistor probes. Thermistors are characterized by their time constants, t_c, which are defined as the time that it takes for a thermistor to reach 63% of its final value when subjected to a temperature pulse [8]. This is directly analogous to time constants for many other electrical and mechanical devices. It will take five times the t_c for a thermistor to reach 99.3% of its final value. Time constants for thermistors vary from 0.1 to 100 s or more. A reasonable device should be designed so that the time constant of the thermistor is not the limiting factor in the analysis.

Major factors which influence the time constant of a thermistor are its mass, geometric configuration, and mode of encapsulation. The fastest thermistors are small bare beads. Beads encapsulated in thin glass have time constants on the order of 1 to 10 s. Other forms such as rods and washers have much longer time constants.

E. Noise and Temperature Resolution

Although many claims of very sensitive thermal measurements have been made, Bowers and Carr [10] derived practical limits for thermochemical measurements using thermistors in Wheatstone bridges. They recognize three major sources of noise which contribute to the ultimate resolution of a thermistor. These are

a. Sources which are independent of the voltage applied to the Wheatstone bridge such as pickup noise from fluorescent lights and electric motors, Johnson noise due to the Brownian motion of electrons, and flicker noise which is proportional to the reciprocal of the frequency, $1/f$.
b. Sources of noise dependent upon the voltage applied to the Wheatstone bridge, E_b, with thermal inhomogeneity of the solution being one example.
c. Sources of noise which vary with the third power of the applied voltage and are ascribed to self-heating effects. The third power relationship comes from the combination of Joule heating

of the thermistor itself (proportional to E_b^2) and irregularities in the flow of solution by the thermistor (proportional to E_b as in b) which alter heat dissipation from the thermistor to the solution.

The total noise expressed as a variation in temperature may be expressed as

$$\Delta T = \sqrt{\frac{A}{E_b^2} + B + CE_b^4} \tag{9.4}$$

where A, B, and C are constants related to the three types of noise above respectively, and E_b is the applied bridge potential. This equation predicts that there exists an optimum bride potential (Eq. (9.5)) and that the temperature resolution has a minumum value given by Eq. (9.6) [10].

$$E_{b,opt} = \left(\frac{A}{2C}\right)^{1/6} \tag{9.5}$$

$$\Delta T_{min} = B + \frac{3A^{2/3}C^{1/3}}{2} \tag{9.6}$$

The authors experimentally tested a variety of the parameters thought to make up the total noise to get empirical values for the constants A, B, and C. From several plausible experimental arrangements and the magnitude of the constants A, B, and C, they found that the minimum observable temperature change is in the range of 2 to 10 $\mu°C$. They also recommended using a thermistor with the largest possible resistance. This latter recommendation is limited by the necessity that the measuring device have an input impedance which is at least 100 times that of the bridge impedance.

III. THERMISTOR PROBES

The basic idea of any probe is that after proper calibration, it can be placed in an analyte mixture and after a short time, a meter, a strip chart recorder or a computer reading would indicate the concentration of a particular species. To be viable, this probe would have to produce reliably reproducible results at a relatively rapid rate for a reasonably long lifetime.

A potentiometric probe (e.g., glass electrode) works because of a chemical interaction between the analyte of interest and the probe itself which causes a potential to develop. A thermistor probe must be designed to produce a chemical reaction near the thermistor which will cause a temperature change that can be measured. Unlike the potential of a glass electrode, the thermistor device needs a continuing reaction to operate. In this sense, it is closer to an amperometric detector than a potentiometric detector. In order to have a long lasting probe, it is reasonable that the thermistor should be a catalytic site. To date the most attractive catalysts have been enzymes because of the interest of the biomedical community. Synthetic materials which mimic enzymes may be even better candidates in the future if they prove to be more stable.

A. Thermal Enzyme Probe

The first operational thermistor probe for enzymatic substrate analysis was reported by Cooney et al. in 1974 [11]. The device was called a thermal enzyme probe, and consisted of an active thermistor, coated with an enzyme, and a reference thermistor. In order to immobilize the enzyme, glutaraldehyde crosslinking or simple adhesives were used. The reference thermistor could be uncoated or, to balance thermal conductivity, it could be coated with an inert protein such as bovine serum albumin (BSA). Electronically these comprised the measuring thermistors of a differential setup of the Wheatstone bridge described earlier. This pair of thermistors was then inserted into an insulated container into which the sample would be added. An outer thermostat bath surrounded the apparatus and stirring was provided with a magnetic stirring bar. The apparatus [12] is diagrammed in Figure 2. When the enzyme substrate was added to the mixture, relatively large temperature decreases were observed before the steady-state temperature increase of the enzymatic reaction itself. Typical temperature increases were on the order of 0.1 to 1.0 m°C [12,13].

In describing this probe, Weaver et al. [12] put together a straightforward diffusion model of the process. Their analysis is that the reaction rate at the thermistor surface is controlled by the diffusion of substrate from the bulk solution. Heat generated as a consequence of the reaction velocity and the heat of reaction, ΔH, also diffuses away from the thermistor surface with a known flux. The result is that the temperature change observed will follow the simplified expression

$$\Delta T = \frac{D_s \Delta H C_s}{K} \qquad (9.7)$$

where D_s and C_s represent the diffusion constant and bulk concentration of the substrate, K is the thermal conductivity of the un-

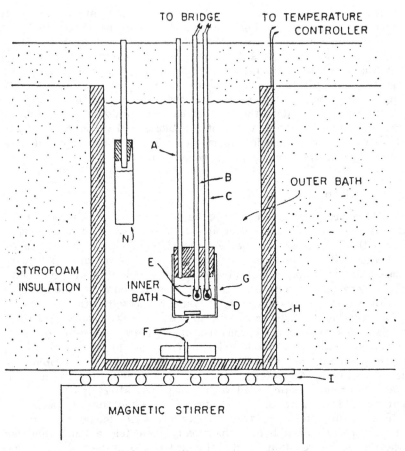

Figure 2   The thermal enzyme probe stirred-bath configuration used for thermal response measurements: (A) sample injection tube, (B,C) stainless steel thermistor tubes, (D,E) thermistors, (F) magnetic stirring bars, (G) glass inner bath wall, (H) brass outer bath wall, (I) cooling coil, (N) temporary storage reservoir for concentrated substrate prior to injection into the inner bath.   (From Ref. 12 with permission.)

stirred layer just around the thermistor, and ΔH is the heat of reaction [12].

Reactions of trypsin, hexokinase, and glucose oxidase were
studied using this apparatus [11,12,13]. A linear increase in temperature with substrate concentration is found from approximately 1
to 4 m$\underline{M}$ for hexokinase and 0.1 to 2 m$\underline{M}$ for glucose oxidase. The
authors do note changes in the apparent Michaelis constant K_m be-
tween the immobilized and soluble enzyme forms. These differences
are taken to indicate a diffusion-controlled reaction across a thin
aqueous layer [12]. This suggests that the reaction is taking place
at the surface of the thermistor where the enzyme is immobilized.

B. Flow-Through Thermal Enzyme Probe

In a later device, Fulton et al. [9] placed their thermal enzyme
probe in a laminar flow cell. The device is illustrated in Figure 3a.
They found the device to be very sensitive to fluctuations in flow
rate, thus precluding the use of mechanical, syringe, and peristaltic pumps. The best solution was to deliver solution from a nitro-
gen-pressurized stainless steel tank. Flow rates less than 1 mL/
min provided laminar flow to reduce the noise of the system [10].
The authors were able to demonstrate the response of this system
to the hydrolysis of urea. They achieved a temperature response
of 0.6 millidegrees at 3×10^{-2} molar urea. Response times to
changes in urea concentrations were on the order of 1–3 min as
shown in Figure 3b.

C. Enzyme Bound Thermistor

To solve the problem of low-temperature changes suggested above
[11–13], Tran-Minh and Vallin [14] have constructed a device
called the enzyme-bound thermistor. Their thermistor is surrounded by a semiporous glass jacket as shown in Figure 4. In this
case, the porous jacket impedes free flow of solution and also heat
from the thermistor region. This results in an amplification of the
temperature rise observed. These authors investigated the catalase,
glucose oxidase, and urease enzyme systems.

Of particular interest is their study of the glucose oxidase system. The reaction apparently involves molecular oxygen:

$$\text{Glucose} + O_2 \rightarrow \text{D-gluconic acid} + \tfrac{1}{2} O_2 \tag{9.8}$$

This oxygen is rapidly depleted in normal solutions and becomes
rate limiting. These authors take advantage of the presence of
catalase as an impurity in the glucose oxidase preparation, by add-

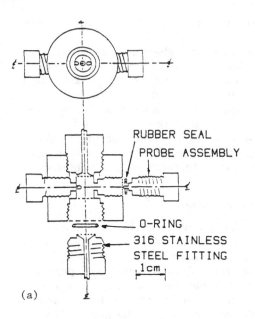

RUBBER SEAL
PROBE ASSEMBLY

O-RING

316 STAINLESS
STEEL FITTING

|1cm|

(a)

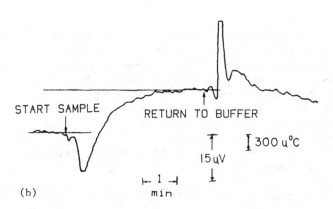

START SAMPLE

RETURN TO BUFFER

$\displaystyle\prod$ 15 μV

$\displaystyle\prod$ 300 u°C

|— 1 —|
min

(b)

Figure 3 (a) Cross-sectional views of the flow through thermal enzyme probe cell. Top: transverse section; bottom: vertical section. (b) Thermal enzyme probe T signal in response to a 7.5 × 10^{-3} $\underline{M}$ urea solution using the sample loop method with a volume flow rate of 0.7 mL min^{-1}. (From Ref. 9 with permission.)

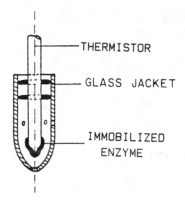

THERMISTOR

GLASS JACKET

IMMOBILIZED
ENZYME

Figure 4 Enzyme-bound thermistor. Note holes in glass jacket to permit limited solution movement. (From Ref. 14 with permission.)

ing H_2O_2 to keep the level of oxygen in the solution high. This resulted in a dramatic improvement in the dynamic range of the analysis as compared to previous results [14].

D. Thermistor Enzyme Probe

Another probe which presented unique aspects was devised by Rich et al. and was called the thermistor enzyme probe [15]. In previous work it had been observed that urease adsorbed to mercury retained some enzymatic activity [16]. A probe was constructed which consisted of a thermistor surrounded by a small bead of mercury to which urease was adsorbed. A reference probe of the same proportions was also used in this differential temperature measurement as described previously. A diagram of the probe is given in Figure 5. The difference between this probe and the others is the mercury metal, which the authors state acts both as an immobilization matrix and as an efficient heat sink because of its high thermal conductivity and low heat capacity [2,15].

The authors of this work use basically the same diffusional model as Weaver et al. [12]. While there is no change in the assumption that substrate must diffuse to the surface of the probe, there is a difference in the diffusion of heat from the enzyme layer. Heat flow will be partitioned between the aqueous system and the mercury. Due to the greater thermal conductivity and lower specific heat of mercury, the aqueous solution now acts as an insulator keeping the heat near the thermistor. The result is a larger temperature change which can then be more accurately and precisely monitored by the thermistor. This gave a response to urea which

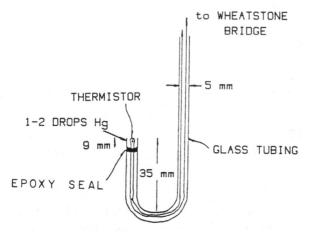

Figure 5 Construction of the thermistor enzyme probe. (From Ref. 15 with permission.)

was approximately 10 times greater than the response observed by Weaver et al. [9].

E. Additional Thermal Probes

Marconi has described a thermistor enzyme probe where the enzymes (glucose oxidase and catalase) were first immobilized on cellulose triacetate fibers, which were then wrapped around the thermistor. As a continuous glucose monitor, it demonstrated linear behavior up to 1 g/L glucose (5 mM) [17].

A last device in this category was developed by Pennington [18, 19]. This involved the coupling of an enzyme to a Peltier device, which is a thermocouple junction through which electrical current is passed. Used in the reverse mode, it acts as a thermocouple and detects temperature changes. Peroxidase was one enzyme used to test this device. Analysis of 10 μL samples of 2×10^{-2} molar peroxide were readily achieved using this instrument.

IV. THE ENZYME THERMISTOR

As opposed to the probe type of thermistor device, an instrument named the enzyme thermistor was developed by Mosbach and Danielsson in 1974 [20]. A prototype constructed around a heat-flow microcalorimeter was reported in 1973 [21]. The enzyme thermistor has its roots in the device described by Priestly et al. [22] where

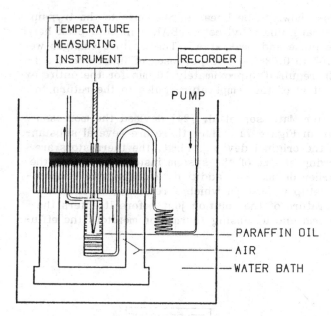

Figure 6 The first enzyme thermistor design. From the pump, the initial coil is a heat exchanger to thermostat the solution. The second coil contains the glass-bound enzyme, and it surrounds the thermistor. The entire apparatus is immersed in a water bath at 27°C. (From Ref. 20 with permission.)

two solutions are brought together in a flow system to react and the extent of that reaction is determined thermometrically. The enzyme thermistor is actually a simpler device than Priestly's, since there is only one solution containing all of the reactants, and the problems of complete and efficient mixing of reactant streams are avoided. The enzyme thermistor device simply provides a column which contains the appropriate catalyst for the reaction (usually an immobilized enzyme) and a postcolumn thermistor to monitor the temperature of the exiting solution.

The first reported enzyme thermistor [20] was designed as shown in Figure 6, to contain a thermistor in paraffin oil surrounded by a coil of tubing cemented in place to form a cup with a diameter of 4 mm. The tubing itself had a volume of 180 µL and was filled with approximately 100 mg of glass beads to which enzyme was attached. When substrate was pumped through the tubing a chemical reaction, catalyzed by the enzyme was sufficient to cause a temperature change noted by the thermistor. The first enzyme systems studied by this device were trypsin and apyrase. A similar device was reported at the same time by Canning and Carr [23].

This device was shown to be linear in response to the trypsin substrate benzoyl-L-arginine ethyl ester (BAEE) in both peak height of the temperature pulse and peak area. Temperature changes were on the order of 0.02 to 0.08°C. For purposes of illustration, a 1-mL sample of BAEE required approximately 10 min for the entire experiment from the start of the temperature pulse to the return to the baseline [20].

Shortly thereafter Matiasson et al. [24] devised the more sensitive device shown in Figure 7a. Here there are several substantive changes from the original device. First, the thermistors are directly in the flowing stream of the system instead of outside the coil as in their earlier device. In addition, a differential temperature measurement setup is used in which a reference thermistor monitors the temperature of the solution just before it enters the enzyme-packed column and a sensing thermistor monitors the efflu-

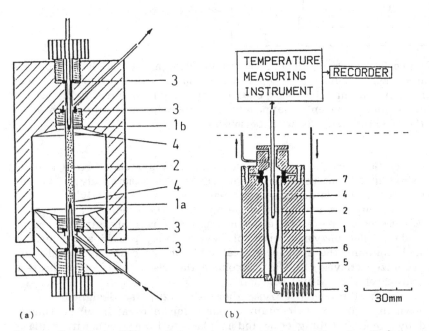

Figure 7 (a) A differential enzyme thermistor unit. 1a, Reference thermistor; 1b, sensing thermistor; 2, Teflon column containing the immobilized enzyme; 3, O-rings; 4, porous polyethylene disks. (b) Single thermistor unit. 1, Microcolumn containing the immobilized enzyme preparation; 2, steel tube with thermistor; 3, heat exchanger; 4, Perspex cylinder; 5, water jacket; 6, air space; 7, O-rings.

ent. These are arranged in a differential mode as described earlier so that the Wheatstone bridge unbalance potential is proportional to the difference in temperature between the two thermistors. A second design, using only one thermistor, was developed to be used in systems where temperature changes were relatively large (Figure 7b). Again, a similar differential temperature measuring device was developed at the same time by Bowers et al. [25,26].

With the dual thermistor apparatus the authors achieve better baseline stability, greater efficiency in measuring a larger fraction of the heat produced and higher flow rates could be used. Temperature changes as low as 10^{-4}°C are reported with maximum temperature changes of 0.2°.

A property called unspecific heat was observed with this device [24]; that is, a temperature pulse is observed which is not related to the enzymatic reaction but to some other process of unknown nature. Figure 8 illustrates the nature of unspecific heat effects. In order to deal with the problem of unspecific heat, these workers developed the split-flow enzyme thermistor [27]. Here the sample stream is divided equally between two columns as shown in Figure 9. One column contains active enzyme as above, while the other column contains the inert support material. Each column also has a reference and sensing thermistor and measurements could be made by

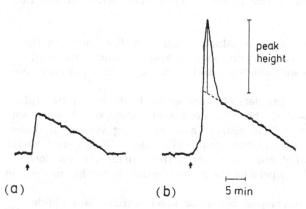

peak
height

(a) (b) 5 min

Figure 8 (a) A typical response curve of "unspecific heat" from the enzyme-thermistor (catalase-glass) using diluted milk with no precolumns. (b) A typical response curve upon applying the diluted milk sample via precolumns with lactase and glucose oxidase. The dotted line is the baseline used when calculating the peak height. (From Ref. 24 with permission.)

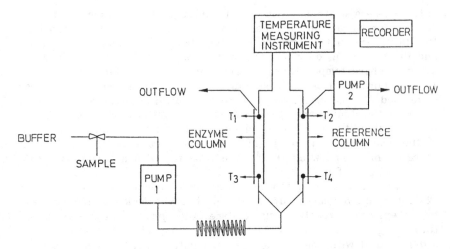

Figure 9 Experimental arrangement of the split-flow enzyme thermistor. P_1 and P_2 are peristaltic pumps. T represents the four
thermistors which may be utilized in the different combinations discussed in the text. (From Ref. 27 with permission.)

a. The sensing thermistor on the active column alone
b. The differential mode on the active column which is identical to
    the device above
c. Measuring the difference in the sensing thermistors of the two
    columns

In addition, an extra pump is added to equalize flow rates in the
two columns. This arrangement made it easy to study the origins
of the "unspecific heat" which caused problems in the previous devices.

In addition to the problem of "unspecific heat" which the split-
flow device helped solve, these authors also developed chemical amplification techniques. This simply involved using several enzymes
so that a sequence of reactions could take place. The result would
be more heat generated and therefore better sensitivity and lower
detection limits [24]. Specific reactions are given in the section on
applications.

Another device developed by these same authors is a single
thermistor device where the thermistor is placed in a chamber containing the immobilized enzyme [3]. The larger volume of this
chamber tends to dampen fluctuations in temperature because of its
volume, yet provide a large amount of enzyme for complete conversion of substrate. The authors feel that its simplicity is advanta-

geous for cases where the sensitivity of the differential mode are not needed. The design is the same as in Figure 7b.

Sensitivity was excellent in the above devices where the differential measurements could readily be made in the 1 to 20 millidegree range. Linearity for the analysis of penicillin G was observed from less than 1 μmol to over 150 μmol of sample [28]. In these two devices, each peak required five minutes before the return to the baseline and recoveries were reported to be ±2% for an analysis [28].

In order to provide a rapid analytical tool, for a variety of substances, Mattiasson and Borrebaeck [29] developed a reversible immobilization column. In this apparatus, the column material contained immobilized lectins which would tightly bind various enzymes such as glucose oxidase and catalase. These enzymes could be added to the column or removed with a wash of 0.2 $\underline{M}$ glycine-HCl, pH 2.2. The analyst could then determine glucose in one sample, wash off the glucose oxidase, add the catalase to the reactor, and then do an analysis of hydrogen peroxide in a relatively short length of time.

V. APPLICATIONS OF THERMAL PROBES

All of the applications using thermistor-based devices involve enzymes. The heat of the reaction catalyzed by these enzymes defines the ultimate sensitivity of each individual analysis. Table 1 is a compilation of the most commonly used enzymes and their heats of reaction. Table 2 is a short compilation of buffer heats of proton ionization based mainly on the data of McGlothlin and Jordan. Since most enzymatic reactions are run in buffered media and involve the consumption or production of hydrogen ions, their contribution is important. Changing the buffer can greatly alter the heat output of a reaction and, therefore, the ultimate sensitivity and detection limits.

There are a variety of areas in which enzymatic analysis is an important component. Two major areas are fermentation and process control and clinical analysis [40].

A. Fermentation Analysis and Process Control

With the development of new biotechnologies, and the interest in biosynthetic processes, there is renewed interest in the maintenance and control of fermentation vats. Chemical analysis of fermentation broths is essential in producing a product which has the greatest yield and best purity. It seems reasonable that a biochemically oriented analysis should be considered for this task. Of particular

Table 1 Some Heats for Reactions Catalyzed by the
Following Enzymes

Enzyme	Heat of reaction (kJ/mol)	References
Catalase	− 100.4	30
Cholesterol oxidase	− 52.9	31
Glucose oxidase	− 80	32
Hexokinase	− 27.6	33
Lactate dehydrogenase	− 62.1	31,34
Trypsin	− 27.8	35
Urease	− 24.3	36
Uricase	− 49.1	31
Malate dehydrogenase	− 89.5	37

[a]Heats do not include buffer protonation.

interest in the biochemically related analysis is that no possibly
harmful chemical species need be used in the process, and the one
essential reagent is itself immobilized within the measuring instrument. A particularly good reason for using thermistor enzyme
probes and the enzyme thermistor apparatus is the nature of the
sample itself. The complex mixture of large molecules and cellular
organisms is one which would require considerable pretreatment for
chromatography. Some separation would certainly be needed for all
but the most simple spectroscopic analysis. Also, in some cases,
the information is needed very rapidly or continuously. A small
probe which reacts with some component of the fermentation broth,
in precisely the same way the organisms do, would be of great value. This is precisely what the thermistor enzyme probes do and it
is also a function of the enzyme thermistor.

   To date there have been several studies of these thermal devices which indicate that they can be used successfully for process
control. Previously, the use of lactase to determine lactose in milk
was mentioned [24] as a clinical analysis, but it is obviously also an
important component in food processing. One of the problems observed with the lactase system is the clogging of the reactor with
solids. This may be solved by prior filtration, dilution by a factor
of 10 or more or by using a hollow fiber reactor. Two antibiotics

Table 2 Heats of Proton Ionization of Some Common
Biological Buffers

Buffer name	pK	Heat of reaction[a] (kJ/mol)
MES	6.08	12.7 ± 0.08
Bis-tris	6.41	29.2 ± 0.17
ACES	6.65	30 ± 0.4
ADA	6.75	11.5 ± 0.38
MOPS	6.76	19.0 ± 0.17
PIPES	6.79	8.7 ± 0.21
BES	6.92	23.1 ± 0.25
Phosphate[b]	7.21	4.74
HEPES	7.24	16.4 ± 0.17
TES	7.34	29.2 ± 0.13
Ethyl glycinate	7.57	46 3 ± 0.08
Glycinamide	7.73	44.8 ± 0.4
HEPPS	7.82	17.9 ± 0.29
Tricine	8.00	30.5 ± 0.21
Tris	8.03	54.4 ± 0.4
Glycylglycine	8.21	44.1 ± 0.13
Bicine	8.31	26.2 ± 0.25
TAPS	8.34	40.1 ± 0.21
N,N-Dimethylglycine	9.95	31.5 ± 0.29
CAPS	10.35	48.5 ± 0.8

[a]All of the above data from Ref. 38 except as noted.
[b]From Ref. 39.

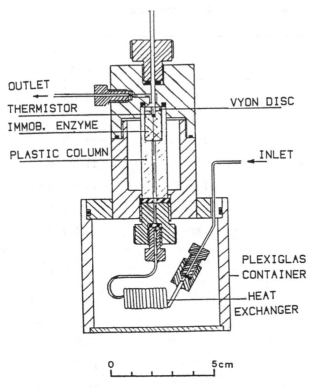

OUTLET

THERMISTOR

IMMOB. ENZYME

PLASTIC COLUMN

VYON DISC

INLET

PLEXIGLAS
CONTAINER

HEAT
EXCHANGER

0   5cm

Figure 10 Schematic drawing of the enzyme thermistor unit used to
monitor and control the lactase enzyme reactor. (From Ref. 47 with
permission.)

have been well studied, penicillin G [41–43] and cephalosporin [41].
Mosbach and Danielsson [4] also report the use of the enzyme thermistor for the analysis of sucrose, galactose, cellobiose, and ethanol,
all of which may be important components of fermentation broths.

Specific applications to fermentation systems are also found in
several papers [47,44–48]. The enzyme thermistor was linked to
an enzyme column to control a simulated industrial reactor [46,47].
The design was such that the effluent from the enzyme reactor column was diluted and pumped to the enzyme thermistor. The response of the enzyme thermistor were then fed to a pump controller
for the enzyme reactor. In this case it was found that the enzyme
reactor, which contained lactase, could be held under optimum conditions by the enzyme thermistor and the feedback circuit. Figure
10 illustrates the enzyme thermistor apparatus, and Figure 11 illustrates the entire apparatus for controlling a lactase column.

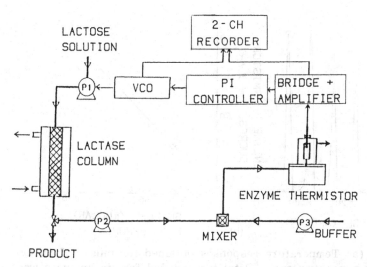

Figure 11   Reactor system controlled by the enzyme thermistor.   Effluent from lactase column is pumped by P_2 and diluted with buffer pumped by P_3 to the enzyme thermistor.   Signal from bridge amplifier is fed back through a voltage-controlled oscillator (VCO) to control the flow of lactose into the column via pump P_1.   (From Ref. 47 with permission.)

Another reactor system was tested to control the sucrose concentration of a fermentation vat [45,48,49].   In this case, a pump withdraws fluid from the fermentation vat and pumps it, after diluting 9:1 with buffer, through the enzyme thermistor.   The response of the enzyme thermistor, which uses immobilized yeast, controls a proportional controller which, in turn, controls the speed of a pump which is adding more sucrose to the fermentation vat.   Manual analyses of the fermentation vat by liquid chromatography show that the enzyme thermistor is faithfully following the concentration of sucrose, as shown in Figure 12.   The control of sucrose in the fermentation vat is shown in Figure 13, where the pump rate is inversely proportional to sucrose concentration until the set point is intentionally doubled at 4 h.

Noncontinuous analyses of fermentation broths for penicillin G have been described [28].   The useful linear concentration range was from 0.1 to 100 m$\underline{M}$ penicillin.   Comparing the results of the enzyme thermistor to a conventional analysis gave a correlation coefficient of 0.997.   Reactors using enzymes immobilized on glass beads and on nylon tubing were also compared.

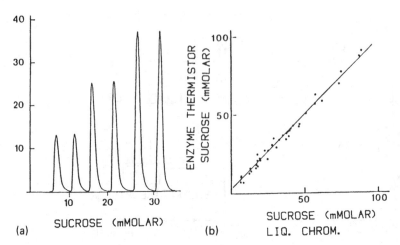

(a) SUCROSE (mMOLAR)

(b) SUCROSE (mMOLAR) LIQ. CHROM.

Figure 12   (a) Temperature responses obtained for pulses of different sucrose concentrations.   Total time needed for six analyses was 30 min.   (b) Correlation between sucrose concentrations measured by the enzyme thermistor and an independent liquid chromatography method.   r = 0.992; n = 39.   (From Ref. 59 with permission.)

An enzyme thermistor was also used to monitor the glucose concentration in a fuel cell [44].   This fuel cell was designed to use the electrochemical potential of <u>Candida</u> <u>utilis</u> at an electrode surface to estimate its concentration in the fermentation broth.

B.   Clinical Analysis

While process control and evaluation systems have used some enzyme systems, many more are reported for clinical analyses and may be adaptable for process control.   These are briefly enumerated below.

It has been shown by several investigators that the use of the enzyme thermistor is suitable for determining urea with urease in the range of 10^{-5} to 0.5 mol/L [9,15,23,25,32,36,41,50−54].   Both the thermistor probes and the enzyme thermistor systems have been utilized.

Glucose and hydrogen peroxide are often test systems [26,32, 41, 52−57].   Several enzymes are used, including glucose oxidase, catalase, and hexokinase.

The study of cholesterol and its esters in the enzyme thermistor system was undertaken by Danielsson et al. [24,41,58].   They found that very good results could be obtained for standard cholesterol solutions.   It was, however, necessary to extract cholesterol from serum with hexane before analysis.

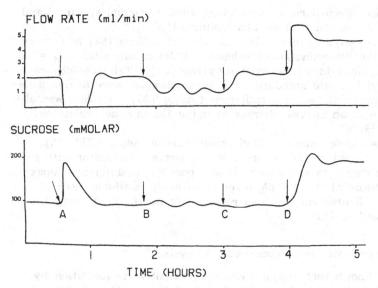

Figure 13 Response of the control system to various disturbances.
The upper curve shows the flow rate of the sucrose pump, while
the lower curve is the thermogram of the enzyme thermistor. From
point A to C the sucrose concentration in the fermenter was set to
100 mM, and in case D it was set to 200 mM. At A, sucrose was
added to increase the sucrose concentration to 200 mM at the arrow.
At B the sucrose concentration of the solution being fed into the
fermenter was changed from 250 to 500 mM; at point C it was
changed back again. At point D the set point was changed to 200
mM. (From Ref. 46 with permission.)

Sugar analysis [41,59] includes lactose [58] which was analyzed
by using a mixture of lactase and glucose oxidase. Galactose was
determined using the enzyme D-galactose oxidase, and sucrose was
determined by hydrolysis to glucose and fructose using invertase.
Cellobiose was hydrolyzed into glucose with glucosidase which then
reacted with glucose oxidase. Ascorbic acid (0.05—0.6 mM) was determined by Mattiasson and Danielsson with the enzyme thermistor
using L-ascorbate oxidase [41].

The analysis of pyruvate with immobilized lactate dehydrogenase
was reported by Schmidt et al. [32]. Lactate was detected by the
enzyme thermistor [60] used as an L.C. detector. Scheller et al.
used a technique called "substrate recycling" in order to increase
sensitivity of the analysis for pyruvate or lactate [61].

Drugs such as penicillin G [42,52], ampicillin [42], and penicillin VK [42] have been studied using thermal methods [41]. Similar

studies have been done on β-lactams, such as cephalosporin C and penicillin V coupled with an FIA system [41,62].

Alcohol dehydrogenase [56] and alcohol oxidase [63] have been used for the determination of ethanol. Calcium and nitrite have been determined in a flow reactor system [64]. Triglycerides such as glyceryl trioleate standards and serum samples were analyzed using the enzyme lipase in a split-flow reactor [65]. Oxalate was also determined in an enzyme thermistor using the enzyme oxalate oxidase [53,56,66].

Thermometric enzyme-linked immunosorbant assay (TELISA) [51,43,67] combines the features of the enzyme thermistor with immunochemistry. In one study, it was possible to determine human serum albumin at the 0.1 pM level in an analysis which takes 11 min [51]. Gentamicin was also reported to be analyzed using the TELISA method [43].

C. Environmental and Trace-Metal Analysis

It is well known that enzymes are very sensitive to inhibition by heavy metals and, at times, their own products [68]. In many instances their inhibition constants, K_i, are in the micromolar range or less. For a reversible inhibitor, the K_i represents the concentration of the heavy metal at which the enzymatic activity is reduced by 50% [69]. In a study in 1975, Baldridge and Jespersen [70] showed how the inhibition of the enzyme urease could be used to analyze for low levels of lead using direct injection enthalpimetry. Application of this to the enzyme thermistor was done by Mattiasson et al. [71]. Use of enzyme inhibition is very attractive since it exhibits great sensitivity. However, it is limited to concentrations approximately one order of magnitude above and below the K_i value. In many instances, commonly used buffers will interfere by precipitating the heavy metals to levels well below the K_i. Last, it is common to find several heavy metals mixed together, and thus the results are not selective for a specific metal. Yet, as a screening tool, it has excellent prospects.

Cyanide was analyzed using the enzyme thermistor with immobilized rhodanase and injectase [72]. These two enzymes utilize cyanide as one substrate with the other substrates being thiosulfate and L-cysteine, respectively. Concentrations from less than 0.05 mM up to 1.0 mM produced a linear calibration curve. The response stability of this device was such that 80% of its activity was retained after more than 250 hr. Finally, it was reported that the insecticide parathion was analyzed using the enzyme acetylcholinesterase in an enzyme thermistor [4]. In that work, it was found that the detection limit was 5×10^{-6} mol/L.

VI. CONCLUSIONS

It is obvious that thermal methods have been well studied under a variety of conditions. Thermal probes are theoretically very attractive but have yet to be regularly used in chemical analysis. Flow devices, like the enzyme thermistor, seem to be much more successful in applications in the industrial area. The enzyme thermistor is selective, responds relatively rapidly, and has been shown to be long lasting. Since a thermistor-based analytical device will not add extraneous chemicals to a system, they are very attractive to sensitive biochemical applications.

REFERENCES

1. Schifreen, R. S., in Biochemical and Clinical Applications of Thermometric and Thermal Analysis, vol. 12, part b (N. D. Jespersen, ed.), Elsevier, New York (1982), pp. 53–74.
2. Ianniello, R. M. and Jespersen, N. D., in Biochemical and Clinical Applications of Thermometric and Thermal Analysis, vol. 12, part b (N. D. Jespersen, ed.), Elsevier, New York (1982), pp. 75–101.
3. Danielsson, B. and Mosbach, K., in Methods in Enzymology, vol. 44 (K. Mosback, ed.), Academic Press, New York (1976), pp. 667–676.
4. Mosbach, K. and Danielsson, B., Anal. Chem., 53, 83A (1981).
5. Mosbach, K., Danielsson, B., and Mattiasson, B., Biochem. Soc. Trans., 7, 11 (1979).
6. Grime, K. J., in Ref. 1, pp. 136–175.
7. Schifreen, R. S., Analytical Solution Calorimetry (J. K. Grime, ed.), Wiley, New York (1985), pp. 97–135.
8. Anonymous, in Thermistor Manual, Fenwal Electronics, Clifton, NJ (1974), pp. 1–13.
9. Fulton, S. P., Cooney, C. L., and Weaver, J. C., Anal. Chem., 52, 505 (1980).
10. Bowers, L. and Carr, P., Thermochim. Acta., 10, 129 (1974).
11. Cooney, C. L., Weaver, J. C., Tannenbaum, S. R., Faller, D. V., and Shields, A., Enzyme Engineering, vol. 2 (E. K. Pye and L. B. Wingard, Jr., eds.), Plenum Press, New York (1974), pp. 411–417.
12. Weaver, J. C., Cooney, C. L., Fulton, S. P., Schuler, P., and Tannenbaum, S. R., Biochim. Biophys. Acta, 452, 285 (1976).
13. Cooney, C. L., Weaver, J. C., Fulton, J. C., and Tannenbaum, S. R., in Ref. 11, vol. 3, pp. 431–436.

14. Tran-Minh, C. and Vallin, D., Anal. Chem., 50, 1874 (1978).
15. Rich, S., Ianniello, R. M., and Jespersen, N. D., Anal. Chem., 51, 204 (1979).
16. Santhanam, K., Jespersen, N. D., and Bard, A., J. Am. Chem. Soc., 99, 274 (1977).
17. Marconi, W., in Enzyme Engineering, Vol. 4 (G. B. Brown, G. Manecke, and L. B. Wingard, Jr., eds.), Plenum Press, New York (1978), pp. 179—186.
18. Pennington, S. N., Enzyme Technol. Digest, 3, 105 (1974).
19. Pennington, S. N., Anal. Biochem., 72, 230 (1976).
20. Mossbach, K. and Danielsson, B., Biochim. Biophys. Acta, 364, 140 (1974).
21. Johannson, I., Lundberg, J., Mattiasson, B., and Mosbach, K., Biochim. Biophys. Acta, 304, 217 (1973).
22. Priestly, P. T., Sebborn, W. S., and Selmon, R. F., Analyst, 90, 589 (1965).
23. Canning, L. M., Jr. and Carr, P. W., Anal. Letts., 8, 359 (1975).
24. Mattiasson, B., Danielsson, B., and Mosbach, K., Anal. Letts., 9, 217 (1976).
25. Bowers, L. D., Canning, L. M., Jr., Sayers, C. N., and Carr, P. W., Clin. Chem., 22, 1314 (1976).
26. Bowers, L. D. and Carr, P. W., Clin. Chem., 22, 1427 (1976).
27. Mattiasson, B., Danielsson, B., and Mosbach, K., Anal. Letts., 9, 867 (1976).
28. Mattiasson, B., Danielsson, B., Winquist, F., Nilsson, H., and Mosbach, K., Appl. Environ. Microbiol., 41, 903 (1981).
29. Mattiasson, B. and Borrebaeck, C., FEBS Letts., 85, 119 (1978).
30. Nelson, D. P. and Kiesow, L. A., Anal. Biochem., 49, 474 (1972).
31. Rehak, N. N. and Young, D. S., Clin. Chem., 24, 1414 (1978).
32. Schmidt, H. L., Krisam, G., and Grenner, G., Biochim. Biophys. Acta, 429, 283 (1976).
33. McGlothlin, C. D. and Jordan, J., Anal. Chem., 47, 786 (1975).
34. Donnovan, L., Barclay, K., Otto, K., and Jespersen, N., Thermochim. Acta, 11, 151 (1975).
35. Brown, H. D., in Biochemical Microcalorimetry (H. D. Brown, ed.), Academic Press, New York (1969), pp. 149—164.
36. Jespersen, N. D., J. Am. Chem. Soc., 97, 1662 (1975).
37. Jespersen, N., Thermochim. Acta, 17, 23 (1976).
38. McGlothlin, C. D. and Jordan, J., Anal. Letts., 9, 245 (1976).

39. Watt, G. D. and Sturtevant, J. M., Biochemistry, 8, 4567 (1969).
40. Weaver, J. C., Cooney, C. L., Tannenbaum, S. R., and Fulton, S. P., in Biomedical Applications of Immobilized Enzymes and Proteins, vol. 2 (T. M. S. Chang, ed.), Plenum Press, New York (1977), pp. 191—205.
41. Danielsson, B., Mattiasson, B., and Mosbach, K., Pure Appl. Chem., 51, 1443 (1979).
42. Grime, J. K. and Tan, B., Anal. Chim. Acta, 107, 319 (1979).
43. Mattiasson, B., Svensson, K., Borrebaeck, C., Jonsson, S., and Kronvall, G., Clin. Chem., 24, 1770 (1978).
44. Miyabayashi, A., Danielsson, B., and Mattiasson, B., Ann. NY Acad. Sci., 501, 555 (1987).
45. Mattiasson, B., Danielsson, B., and Winquist, F., in Enzyme Engineering, vol. 5 (H. H. Weetall and G. P. Royer, eds.), Plenum Press, New York (1980), pp. 251—254.
46. Mattiasson, B., Danielsson, B., Mandenius, C. F., and Winquist, F., Ann. NY Acad. Sci., 369, 295 (1981).
47. Danielsson, B., Mattiasson, B., Karlsson, R., and Winqvist, F., Biotech. and Bioeng., 21, 1749 (1979).
48. Mandenius, C. F., Danielsson, B., and Mattiasson, B., Acta Chim. Scand., B34, 463 (1980).
49. Mattison, B., Larsson, P.-O., and Mosbach, K., Nature, 268, 519 (1977).
50. Rehak, N. N., Janes, G., and Yound, D. S., Clin. Chem., 23, 195 (1977).
51. Mattiasson, B., Borrebaeck, C., Sanfridsson, B., and Mosbach, K., Biochim. Biophys. Acta, 483, 221 (1977).
52. Mosbach, K., Danielsson, B., Borgerud, A., and Scott, M., Biochim. Biophys. Acta, 403, 256 (1975).
53. Dannielsson, B. and Mosbach, K., FEBS Letts., 101, 47 (1979).
54. Kirch, P., Danzer, J., Krisam, G., and Schmidt, H.-L., in Ref. 17, pp. 217—218.
55. Danielsson, B., Gadd, K., Mattiasson, B., and Mosbach, K., Clin. Chim. Acta, 81, 163 (1977).
56. Danielsson, B., Ann. NY Acad. Sci., 501, 543 (1987).
57. Adlercreutz, P. and Mattiasson, B., Acta Chim. Scand., B36, 651 (1982).
58. Mattiasson, B., Danielsson, B., and Mosbach, K., in Ref. 17, pp. 213—216.
59. Mattiasson, B. and Danielsson, B., 102, 273 (1982).
60. Danielsson, B., Buelow, L., Lowe, C. R., Satoh, I., and Mosbach, K., Anal. Biochem., 117, 84 (1981).
61. Scheller, F., Siegbahn, N., Danielsson, B., and Mosbach, K., Anal. Chem., 57, 1740 (1985).

62. Decristoforo, G. and Danielsson, B., Anal. Chem., 56, 263 (1984).
63. Guilbault, G. G., Danielsson, B., Mandenius, C. F., and Mosbach, K., Anal. Chem., 55, 1582 (1983).
64. Schifreen, R. S., Miller, C. S., and Carr, P. W., Anal. Chem., 51, 278 (1979).
65. Satoh, I., Danielsson, B., and Mosbach, K., Anal. Chim. Acta, 131, 255 (1981).
66. Winquist, F., Danielsson, B., Malpote, J.-Y., Persson, L., and Larsson, M.-B., Anal. Letts., 18, 573 (1985).
67. Borrebaeck, C., Borjeson, J., and Mattiasson, B., Clin. Chem. Acta, 86, 267 (1978).
68. Hoare, J. P. and Laidler, K., J. Am. Chem. Soc., 72, 2487 (1950).
69. Lehninger, A. L., in Biochemistry, 2nd ed., Worth, New York (1975), pp. 183—216.
70. Baldridge, J. N. and Jespersen, N. D., Anal. Letts., 8, 683 (1975).
71. Mattiasson, B., Danielsson, B., Hermansson, C., and Mosbach, K., FEBS Letts., 85, 203 (1978).
72. Mattiasson, B., Mosbach, K., and Svensson, A., Biotech. and Bioeng., 19, 1643 (1977).

10

Flow Injection Analysis in Bioprocess Control

ARTHUR L. REED   *Control Equipment Corporation, Lowell, Massachusetts*

I. BACKGROUND

Flow injection analysis (FIA) is a relatively novel method of chemical analysis which at present is largely restricted to well-defined laboratory analytical applications. However, the characteristic short analysis time and the analytical flexibility of FIA lend it to potential applications as a process analytical tool. This chapter outlines the operating principles of FIA and describes several examples of the application of the method for on-line bioprocess monitoring.

FIA, as with many other on-line analytical methods, comprises three basic building blocks that must be used in various combinations. First, sample handling and transport: to physically move the sample from the field to the lab and make the physical properties of the sample conducive to the technique (e.g., digesting the sample in acid until it is completely dissolved); second, speciation: to chemically alter the sample to either eliminate interferences or produce another species for detection proportional to the species under scrutiny; and third, detection: a transducer of some type that yields a signal proportional to the occurrence of the species of interest. FIA can be a very flexible analyzer that can be molded to use the three building blocks in any number of complex combinations, or it can be a very rigid and simple analyzer set up to run a specific analysis on specific samples using a minimal combination of the three basic blocks. In either case, the FIA system has advantages as well as disadvantages over other systems presently used for similar data acquisition. One major advantage of using FIA to perform an on-line analysis is that it automatically calibrates itself with standards providing more reliable results. If the detector system is prone to drifting, then constant monitoring with standards is the only way to verify sample data.

II. PRINCIPLES OF OPERATION

In detail, FIAs theory of operation is simple yet elegant [1—8].
FIA can be viewed as a miniature closed process which takes in raw
material (sample or standard) and processes it until it reaches the
detector(s). The resulting output is proportional and specific to
only the species of interest (Figure 1). A liquid sample is drawn
into a sample loop of 1—250 µL, typically. A carrier stream, which
may also be a reagent/indicator, is constantly flowing through a detector. The sample in the loop is transposed into the carrier
stream, which pushes the sample to the detector. Various chemical
and physical steps may occur continuously to the carrier and, thus,
to the sample. Ultimately, the sample reaches the detector, which
responds by giving a gaussian-like peak. Data analysis can utilize
peak height, width, or area. Calculating the relative standard deviation (RSD) after many experiments for a particular analysis is
the best indication as to whether height, width, or area is to be
used. FIA differs from other techniques in that the sample may not
have reached chemical and/or physical equilibrium with the reagent/
indicators upon detection. As a matter of course, the transient response is typically not at equilibrium. This is not of concern,
since the flow/rates and sample volumes are so precisely maintained

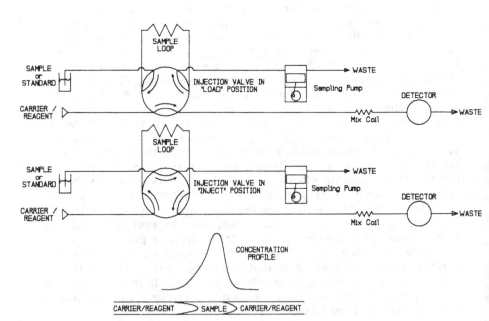

Figure 1   Schematic representation of a simple FIA system.

that the signals generated at the detector will be repeatable for a given concentration of species. The only time that this may not work is if chemical development to equilibria is significantly sensitive to any deviation of the system parameters.

There are several alternative automated analysis techniques. These include automatic batch analyzers and segmented flow analyzers (Technicon); however, neither achieves real-time analysis required in many bacterial fermentation applications. Automatic batch analyzers operate by continuously moving a chain of vessels where samples, reagents, mixing, and heat are added automatically before being analyzed by a detector. A laboratory robot operates similarly, except that the robot moves the reaction vessels individually. Segmented flow analyzers are similar to automatic batch analyzers, except the vessels are individual segments of sample/reagent separated by bubbles flowing through a plastic and glass conduit system. In either case the reaction of samples and reagents must reach equilibrium before being detected. The time delay between sampling and detection can be as long as 30 minutes. Typical response time for FIA methods is approximately 30 s due to the fact that FIA does not have to wait for reactions to reach equilibrium. More details on the theory and rules of thumb for designing a FIA manifold are available [1–3].

A FIA system can be made by physically assembling modules together in a way that suits the chemistry. Its basic building blocks are those of analytical chemistry: sample handling, transport, speciation, and detection. Now, we will discuss the individual components that make up an FIA system and how they have been utilized in bioprocess monitoring.

A. Sample Handling

Sample handling in relation to FIA can be accomplished automatically by any physical means (Table 1). A robotic arm can serve the FIA system with prepared samples as it might a liquid chromatography analyzer. More simplistically, an automatic sampler which dips into individual cups of liquid samples would also suffice. To extract a sample from a process loop, a tangential flow filter can provide the FIA with continuous sampling. Clearly, there are many sampling techniques, but bioprocessing needs special considerations.

In order for the bioprocess to be analyzed without breach of its sterility, it is important to sample the broth without permitting chemical flow from the outside. It is just as important that live biologicals do not leave the fermenter prematurely. To solve the challenge, many sampling systems have been developed by Recktenwald's group [9]. These systems use a small turbulent mixing chamber that keeps the cells from blocking the 0.45-μm HVLP Millipore filter.

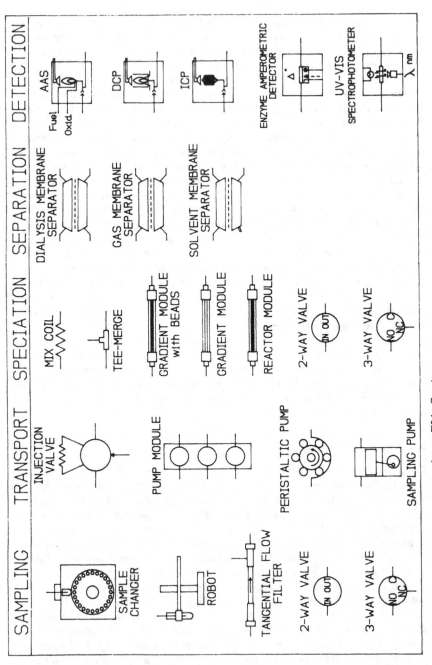

Table 1 Typical Components in a FIA System

This small mixing chamber is constantly recirculating with fermenter broth and cells. The filtrate is pulled from the other side of a Millipore filter. An alternative system, developed by Control Equipment Corporation in cooperation with Worcester Polytechnical Institute [10], utilizes a small piece of a cell-harvesting filter, made by Norton Co., called the Ceraflo filter. This element has a particle size cutoff of 0.2 μm and is a tubular ceramic material. It is installed within a stainless steel housing of very low dead volume and outfitted with silicone or Viton tubing. The flow flux through the inside diameter of the filter keeps cells from blocking the 0.2-μm surface. The filtrate is extracted via a perpendicular 1/16-in stainless steel tube brazed to the housing. A third system, developed by Nikolajsen [11] and associates, takes a two-stage approach. In this sampling system an aliquot of the broth is extracted into a small vessel. Then a small amount of an antimicrobial agent is added to deactivate the cells. The sequence involves filling the sample loop with the now dead sample. The sample is injected into a distilled water carrier which passes over a 0.45-μm Gelman filter. The filtrate side of the filter has a flowing stream that is tied into the rest of the FIA system.

One special consideration when using any sampling on a fermenter is gas infiltration. Where anaerobic fermentations are employed, Viton, PVC, or stainless steel must be used to make connections. Silicon tubing is extremely porous to most gases, including oxygen. These sampling devices now obtain a safe portion of the broth without causing damage to the culture. All of the sampling devices described allow the FIA system to analyze the fermenter in real time with as little as 50 μL of sample. The 50 μL includes one throwaway sample to take care of holdup volume in the sampling device. This small sample has a significant advantage over manual extraction and analysis. It is even superior to a robotic approach.

B. Transport

Transport is a crucial component of FIA (Table 1). The sample needs to be transported by flow through a narrow I.D. tube (sample loop) to fill a very repeatable volume. This volume is then transposed by the injection valve into the carrier stream. The carrier stream transports the sample on to other modules, including the detector(s).

It is important that the design of the sampling system filter out large particles causing a potential blockage problem. Blockage or large carryover can still occur, particularly with proteins. Proteins as well as certain dyes can sometimes have an affinity for the Teflon

tubing used as the conduits in a FIA system. To remedy this situation, an extra wash stream can be switched in after each analysis.
One alternative technique is the addition of a surfactant to the carrier stream. As a rule, these problems do not arise, but it is better to be aware of them rather than discover them. Dissolved gases
in the sample also create a problem during sample transport. These
gases may escape slowly and collect into bubbles. Bubbles usually
mean erratic, disproportional readings from most flow-through detectors. To avoid this, an in-line debubbler made from GORE (registered trademark of W. L. Gore & Associates) tubing can be used.
Another approach, although more expensive, is to use a gas membrane module (Table 1) from an FIA manufacturer.

C. Speciation

Speciation is probably the most significant aspect of chemical analysis, its role in FIA being no exception. Without speciation the signals coming from most detectors just indicate the presence of an
analyte above the background. Detectors usually have some inherent selectivity built in. But if the detector is not specific enough,
it needs help.

    Part of speciation is the chemical and physical manipulation of
the sample and carrier to make them conducive to the detector's
needs. For a detector using immobilized enzyme, the pH is critical.
Extremes in pH can shorten the life of the detector significantly.
By incorporating a buffer into the transport stream, the detector
can operate in its optimal pH range. This is a major advantage
over using the detector directly in the fermenter.

    Some actual examples of speciation in bioprocess monitoring are
given; in the use of biuret [9] and Bradford [9], the reagents
yield a color change when protein is present, which can be detected
readily by a spectrophotometer. The same can be said of the enzymatic color changes using NAD^+ [12]. Molybdovanadate has also
been used with a spectrophotometric detector for phosphate analysis [10]. Because a photodiode has virtually no selectivity, enzymes [11] are used to luminesce only when a particular species is
present.

D. Detectors

Under practical applications there are various detectors mentioned
that are presently being used in FIA for bioprocess monitoring.
Detection, the final building block, is often the most complex. Today, a wide variety of flow-through detectors available are suitable
for FIA. However, even though these detectors have built-in spe-

ciation, they can almost always be enhanced by prior speciation. It is this prior manipulation of the sample combined with the user-independent operation of FIA which results in highly repeatable and accurate results, results that are as of yet unmatched by other forms of detection.

III. PRACTICAL APPLICATIONS IN BIOPROCESSES

A. Monitoring Protein

Introduction

Using a commercial flow injection analyzer (Tecator FIA Star-5020), protein concentration was analyzed in real time during cultivation of Candida boidinii in disruption of Saccharomyces cerevisiae in glass-bead mills [9]. Two methods were tested. The biuret assay worked well for 0.1 to 40 mg/mL protein and was linear. The second employed the Bradford assay. The Bradford assay has been shown to be less susceptible to interferences and very sensitive, but requires a washing of the FIA system between each analysis.

Chemistries

The biuret [13,14] chemistry was adapted to the FIA system. The sample was injected into a carrier of 2% KOH (0.8 mL/min) and proceeded through a mixing coil. The sample volume injected was 100 μL. The stream continued on to a spectrophotometer set at 550 nm.

The Bradford [15] chemistry was installed on the FIA system including a special wash-water flow of 1.5 mL/min. The sample was injected into a carrier of deionized (DI) water (2.8 mL/min), merged with detergent (1.2 mL/min), and continued into a mixing coil. DI water (1.5 mL/min) merged with the outflow of the mixing coil, and finally reagent was merged at 1.5 mL/min and into a spectrophotometer set at 595 nm. The sample volume injected was 35 μL.

Hardware

Sampling devices. The sampling was performed by a "dynamic filtration" device [16] which uses a Millipore HVLP 0.45-μm membrane. This dynamic filter (a stirred volume filter) was connected in-line with the fermenter via a peristaltic pump. The filter assembly is steam-sterilizable. The sampling device had a holdup time of 2 minutes with a filtrate flow rate of 0.34 mL/min.

FIA system. The flow injection analyzer (FIAStar 5020 Tecator) used in this application had full microprocessor control over its

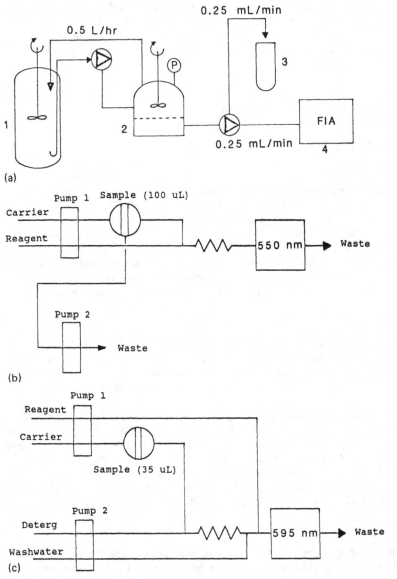

Figure 1 (a) Coupling of sampling device and FIA to the fermenter: 1, fermenter, 2, sampling device; 3, fraction collector; 4, FIA. (b) Flow injection schematic for biuret application. (c) Flow injection for Bradford application. (Figures (a)–(c) from Ref. 9.)

pumps, injection valve, and data reduction. The pumps were small stop/start peristaltic types. The injection valve was a rotary type, which may necessitate rebuilding if many injections per day are required. The built-in microprocessor performed sample data manipulation including automatic standard curve generation during calibration.

Detectors. For the biuret analysis, the spectrophotometer (PM 2 Zeiss) was set at 550 nm. For the Bradford analysis it was set at 595 nm.

Procedure

During the cultivation of Candida boidinii the Bradford assay was used (Figure 1(a),(c)). The sampling device constantly fed the FIA system at 0.25 mL/min. A timer was employed to automatically perform analysis at 6-minute intervals. A fraction collector was used at 0.25 mL/min, such that manual analysis could be performed to show proper correlation with the automated technique.

The analysis of disintegrated Saccharomyces cerevisiae was accomplished by use of the automated biuret method (Figure 1(a),(b)). The S. cerevisiae was disrupted in a model LME20 mill (Netzsch, Selb, FRG). A loop was made by connecting the mill, pump, and a mixing vessel. A secondary loop was then made between the mixing vessel and the sampling device. The filtrate flow rate was maintained at 0.34 mL/min. The FIA system was set to analyze at intervals of approximately 2 minutes.

B. Monitoring β-Glucose and Phosphate Simultaneously

Introduction

The media of recombinant strain of the S. cerevisiae at Worcester Polytechnic Institute (Figure 2(a),(b)) in a small benchtop fermenter was monitored every 5 hr automatically [10]. The sampling was performed via a tangential flow filter and required only about 100 μL total media per analysis cycle. The analysis cycle included a single point calibration as well as one throwaway sample of the media to purge system holdup. The FIA analysis was sequential, utilizing a flow-through immobilized enzyme electrode (β-glucose) and a flow-through spectrophotometer.

Chemistries

β-glucose was analyzed via a special composite membrane stretched over an oxygen electrode (Figure 2(a)). The composite membrane has glucose oxidase entrapped between its layers. When a sample containing β-glucose comes in contact with the membrane,

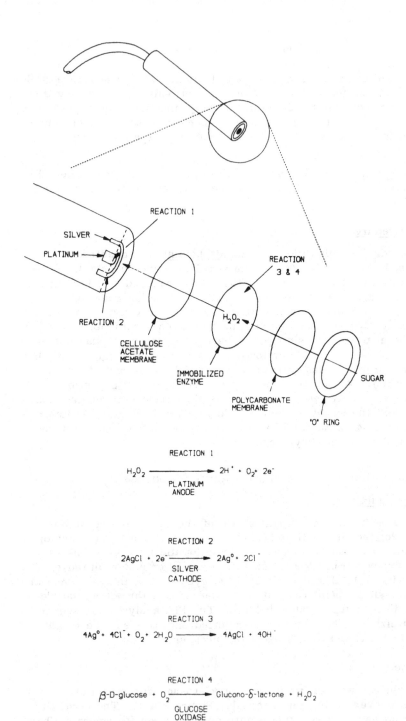

REACTION 1

SILVER

PLATINUM

REACTION 2

CELLULOSE
ACETATE
MEMBRANE

REACTION
3 & 4

H_2O_2

IMMOBILIZED
ENZYME

POLYCARBONATE
MEMBRANE

"O" RING

SUGAR

REACTION 1

$$H_2O_2 \longrightarrow 2H^+ + O_2 + 2e^-$$

PLATINUM
ANODE

REACTION 2

$$2AgCl + 2e^- \longrightarrow 2Ag^0 + 2Cl^-$$

SILVER
CATHODE

REACTION 3

$$4Ag^0 + 4Cl^- + O_2 + 2H_2O \longrightarrow 4AgCl + 4OH^-$$

REACTION 4

$$\beta\text{-D-glucose} + O_2 \longrightarrow \text{Glucono-}\delta\text{-lactone} + H_2O_2$$

GLUCOSE
OXIDASE

Figure 2(a)   Immobilized enzyme electrode detail.

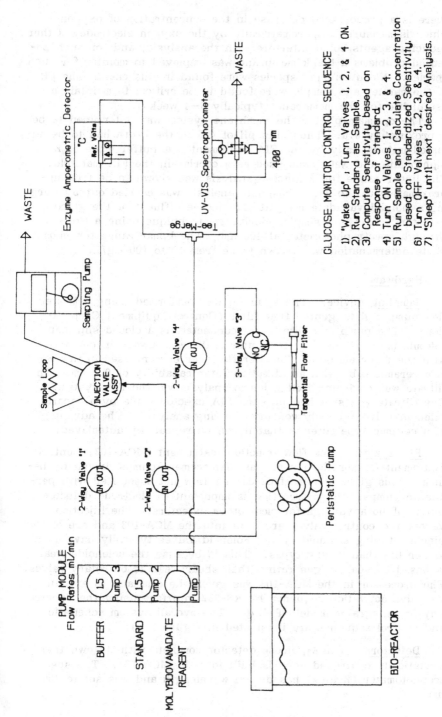

GLUCOSE MONITOR CONTROL SEQUENCE

1) "Wake Up", Turn Valves 1, 2, & 4 ON.
2) Run Standard as Sample.
3) Compute Sensitivity based on Response of Standard
4) Turn ON Valves 1, 2, 3, & 4.
5) Run Sample and Calculate Concentration based on Standardized Sensitivity.
6) Turn OFF Valves 1, 2, 3, & 4.
7) "Sleep" until next desired Analysis.

Figure 2(b)   Glucose and phosphate FIA monitor for bioprocesses.

there is a proportional decrease in the concentration of oxygen
which is measured amperometrically by the oxygen electrode. Other
reducing agents could interfere with the analysis, and for such potential problems a blank membrane was employed to monitor for such
species. No interfering species were found in this case. The pH
of the carrier and sample were found to be critical to maintain a
useful life for the membrane (typically 3–4 weeks).

The optimum pH for the membrane system was determined to be
approximately 6.5. Thus, the pH of 4.5 in the broth had to be adjusted automatically by the FIA equipment. A buffer made of K-
biphthalate and NaOH was used as a carrier in the FIA manifold.
The linear range for β-glucose analysis was shown to be from approximately 0.1 to 2%. Phosphate analysis was carried out by merging a molybdovanadate reagent [17] to the effluent of the glucose
detector. This is a simple colorimetric technique using a flow-
through spectrophotometer at 400 nm. The linear range for phosphate determination was shown to be from 10 to 100 mg/L.

Hardware

Sampling device. The sampling was performed using a Master-
flex pump and tangential flow filter (Control Equipment Corporation). The pump and filter were connected as a closed slip loop
(60 mL/min). Thus, the fermenter media was always in contact
with the filter element. The Ceraflo filter element used in this work
is a ceramic tube with an effective filter capability of 0.2 μm. The
filtrate was cold-sterilized prior to analysis via the FIA system.
The filtrate was switched into the FIA injector valve by a separate
solenoid valve and only consumed during analysis. The advantage
of a ceramic type filter is that it can be repeatedly autoclaved.

FIA system. The flow injection instrument (MCA-103, Control
Equipment Corporation) used gas displacement pumps driven by he-
lium. This gives a pulse-free carrier flow not associated with peristaltic pumps. Pulse-free flow is important in electrode detectors,
but is of no advantage for most other detectors. The injection
valves and control valves are built into the MCA-103 and are of the
pinched-tubing solenoid type. Solenoid valves typically have a
longer life than rotary types. This is because the solenoids wear
is based on seal fatigue rather than shear wear as on rotary valves.
The processor in the MCA-103 was given the command to execute
its valve activation sequence by RS-232 interface using the Spectra-
physics integrator model SP-4270. The overall system schematic
and control sequence are illustrated in Figure 2(b).

Detectors. The β-glucose detector complete with its own thermostatically controlled oven is built into the MCA-103. The spectrophotometer (Kratos) has dial-in wavelength and was set to 400
nm.

C. Monitoring of Formate Dehydrogenase and l-Leucine Dehydrogenase Enzymes

Introduction

On-line analysis of l-leucine dehydrogenase (l-LeuDH) in a filtrate from high-pressure homogenized <u>Bacillus cereus</u>, as well as on-line analysis of formate dehydrogenase (FDH) in filtrate and retentate from <u>Candida boidinii</u> homogenate was performed [12]. Because the analyses were enzymatic, a special thermostatically controlled mixing coil assembly (Figure 3(a)) was made, which yielded $\leq 0.1°C$ deviation. The reaction chemistry used the absorption change at 340 nm when the enzymes combine with NAD^+ to yield NADH. Because the NAD^+ is more expensive than the sample, it was decided to use the sample as the reagent and to inject a diluted amount of NAD^+ as the sample.

Chemistries

For the FDH analysis:

$$HCOO^- + NAD^+ \xrightleftharpoons{FDH} CO_2 + NADH$$

The reaction coil temperature was set at 53°C. This was an optimum temperature for sensitivity and yielded <1% RSD.

For the LeuDH analysis:

$$l\text{-Leucine} + H_2O + NAD^+ \xrightleftharpoons{LeuDH} \text{Ketoleucine} + NH_4^+ + NADH$$

The reaction coil temperature was set at 56°C. This was an optimum temperature for sensitivity and yielded <0.5% RSD.

Hardware

All hardware was the same as described in the section on protein monitoring except for the specially designed thermostatic mixing coil. The mixing coil assembly in this case consists of two coils wound around a metal cylinder through which thermostatically controlled water is pumped. The mixing coil and metal cylinder are enclosed in a polyacrylic box filled with polystyrene beads (Figure 3(a)).

Procedure

<u>FDH analysis (Fig 3(b))</u>. After disintegration of <u>Candida boidinii</u> via potassium phosphate and several hours of stirring, the homogenate was pumped in a closed loop through a pellicon filter

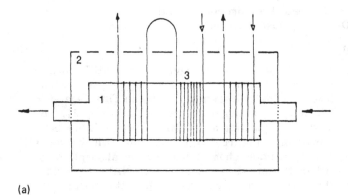

(a)

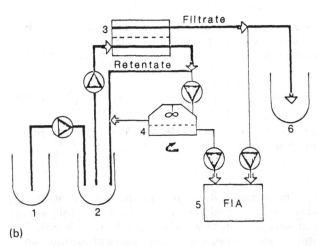

(b)

Figure 3 (a) Temperature control device used for enzyme assays.
Hollow metal cylinder (1) surrounded by a polyacrylic box (2).
Reaction coils of different length (3), which may be connected to
each other. (b) Diafiltration of C. boidinii homogenate. 1, Buffer
added during experiment; 2, suspension vessel; 3, pellicon filter
cassette; 4, sampling device; 5, FIA device; 6, outlet to next
downstream process.

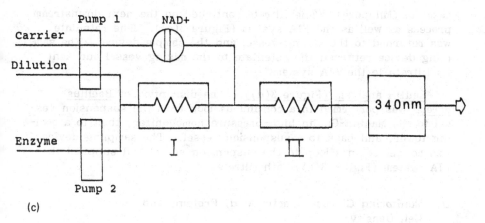

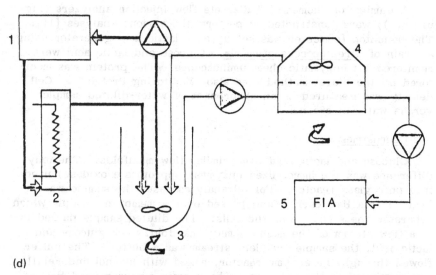

Figure 3 (c) FIA enzyme assay setup. I, Mixing and preheating coil (length/inner diameter 60 cm/0.8 mm). II, Reaction coil (length/inner diameter 120 cm/0.8 mm). Both coils were thermostatically controlled to the enzyme's temperature optimum (FDH: 53°C, LeuDH: 56°C). The NADH concentration was measured spectrophotometrically. (d) Experimental set-up for disruption of Bacillus cereus in a Mauton-Gaulin high pressure homogenizer. 1, Mauton-Gaulin; 2, cooling system; 3, suspension vessel; 4, sampling device; 5, FIA. (Figures (a)-(d) from Ref. 12 with permission from Butterworth.)

cassette (Millipore). The filtrate continued on the next downstream process as well as the FIA system (Figure 3(c)). The retentate was returned to the mixing vessel and the sample device. The sampling device returned the retentate to the mixing vessel and sent its filtrate to the FIA system.

LeuDH analysis (Figure 3(d)). The disruption of Bacillus cereus was performed in a continuous loop from the suspension vessel to the Mauton-Gaulin high-pressure homogenizer, then to a cooling tower, and back to a suspension vessel. The sampling device was connected in a loop to the suspension vessel and supplied the FIA system (Figure 3(c)) with filtrate.

D. Monitoring Glucose, Lactic Acid, Protein, and Cell Density

Introduction

A number of "homemade" discrete flow injection analyzers (Figure 4(a)) were constructed to perform the various analyses [11]. The benchtop fermenter was set up as a lactic acid generator using a strain of Streptococcus cremoris. Glucose and lactic acid were monitored via enzymatic chemiluminescence. The protein was monitored by the biuret method (see also "Monitoring Protein"). Cell density was measured as an absorbance of water-diluted sample versus water at 565 nm.

Chemistries

Glucose and lactic acid used similar flow manifolds. The only difference was the immobilized enzymes (appropriate oxidase) used in a pore glass reactor. For all analyses cases the sample was injected into a H_2O carrier and passed over a membrane module which separated the sample from the cells. The diluted sample passed into a flow stream of the main chemical manifold. For glucose and lactic acid, the sample recipient stream was a buffer. The buffer flowed through the enzyme reactor, mixed with luminol and Fe(III), and continued to the detector. The protein analyzer used H_2O as the sample recipient, which then merged with biuret solution to reach the spectrophotometric detector, set at 565 nm.

Hardware

Sampling device. Rather than a direct on-line sampling device, some work has examined a two-stage approach [11]. The sample was withdrawn from the fermenter and placed in a small sample chamber. Once in the chamber an antimicrobial agent was added to

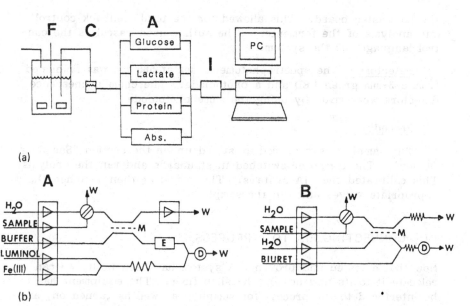

(a)

(b)

Figure 4 (a) Automated system for monitoring fermentation variables. F, fermenter; C, sample chamber; A, analyzers; I, interfacing hardware; PC, personal computer. (b) Flow injection manifolds: (A) for determination of glucose or lactic acid; (B) for determination of protein. W, waste; M, membrane; E, enzyme reactor; D, detector. (From Ref. 11.)

stop cell activity. The sample was then drawn into the sample injection loop and injected into a carrier of distilled water. The carrier passed over a membrane (Metricel Gelman 0.45 μm) where the cells were separated from the media (Figure 4(b)). The media passed through the filter into the recipient stream on the other side of the membrane. The recipient stream carried the media for final analysis.

FIA system (Figure 4(b)). Manifolds were made using 0.5 mm I.D. Teflon tubing and Marpene pump tubing. Two- and three-way PVC connectors were used to facilitate merging. Multipositional and injection valves were MV8 and MV7FPLC (Pharmacia). Three-way solenoid valves were used (NResearch). Peristaltic pumps were controlled via digital-to-analog converters (DAC) and isolation amplifiers by an M24 Olivetti computer. The computer was connected to a LabMaster board (Techmar) which allowed for the control of the valves and pumps. Also, the detectors signals were monitored by the LabMaster board. This allowed for the total feedback control and analysis of the fermenter. The author used Pascal as the control language for the system.

Detectors. The spectrophotometer set at 565 nm was fashioned from a 3-mm green LED and a photodiode. The chemiluminescence detectors were made by using the same photodiode.

Procedure

The sample was prepared as stated under the section "Sampling Device". The computer switched in standards and ran the analysis. This calibrated the FIA systems. The computer then switched the appropriate valves and ran the sample.

IV. CONNECTING FIA TO A PROCESS

Now that a tested and proven FIA system has been built, we can relocate it to its new home: the shop floor. The equipment has to be interfaced to the process for sampling as well as turned on, and adjusted to set parameters, monitored for sample data and diagnosis, and maintained.

Setting the FIA parameters is usually as simple as typing in a few numbers and turning a few knobs. The alternative is to have "canned" software set defaults upon power-up. The FIA parameters to set are reagent pump flow rates, sample loop volume, sampling time, injection time, wash time, and number of samples. Usually the values of the parameters have already been determined and have to be specified only once upon start-up. Once the FIA system is running, the data has to be monitored—specifically, collected, evaluated, and acted upon. The FIA system signal is a peak that is proportional to the species concentration. The peak height is usually all that is needed; it can easily be calculated since most FIA baselines are extremely stable. As a matter of fact, this baseline stability can be used as a diagnostic tool. When a baseline goes awry, it is usually because a reagent has been completely consumed, or else it indicates that the detector is faulty. But, if baseline drift does occur due to a particular nature of the chemistry, the baseline drift can be calculated out of peaks similar to conventional GC drift cancellation. Other diagnostic information is available (e.g., the peak shape). Various deviations in peak shape can occur; they usually indicate abnormalities in the FIA system.

The maintenance of the FIA equipment is probably the major disadvantage of FIA versus other detection systems. More parts usually mean less reliability. However, this is not always the case in FIA systems. This is particularly true when the chemistry of the

FIA system is designed to enhance the selectivity of the detector. Because of this "soft" use of the detector, it typically needs fewer repairs and/or maintenance. Still, there are a few steps that are necessary to maintain a FIA system. The system should probably be cleaned with acids, bases, and/or solvents occasionally if long-term shutdowns occur often. Cleaning should be done if there is an indication of biological growth (due to unsterilized reagents), dirty reagents, or corrosion of parts due to incompatibilities. Most FIA systems are built of glass, Teflon, Kel-f, and Kalrez (registered trademark of E. I. du Pont); these materials are inert to a wide variety of chemicals. The most common maintenance is the refilling of reagents into the pump reservoirs. Reagents should be prepared in a laboratory where they are degassed and filtered.

A. The Economics of FIA

The cost of FIA equipment is evaluated like any other capital investment. The first cost is the capital outlay. The capital expense can range from $1500 for "homemade" manual injection systems to $50,000 for a fully automated multichannel commercial unit. Commercial units that are single-channel and automated can be as low as $7000. The second cost is operational, which includes reagent consumption per analysis, overhead of operator, maintenance, real estate, and power. Because FIA uses very little sample and reagent, as well as reduced operator time, its operational costs are less than virtually any other analysis. Actually, it has been shown [10] that payback on commercial FIA equipment can be less than a month when compared to other techniques.

B. Process FIA in the Future

FIA has come a long way since its early days, but it has a long way to go before it is considered a hardened system for plug-in process control. Better sampling systems need to be designed for extremely harsh environments, and new filter systems are needed that can handle both extremes in chemistry as well as physical properties of the process stream.

Detectors and techniques that have not been used with FIA, which may have significant impact, are multispecies electrodes, x-ray fluorescence, and neutron activation analysis (NAA). The new technology of multispecies electrodes with signal processing may open the door to inexpensive state-variable process control. X-ray fluorescence has an excellent high-speed capability to detect low levels of various elements. But x-ray also has a substantial problem with matrix interference. With FIA the matrix could be somewhat normalized, and thus a high-speed highly specific elemental

analysis could be performed on a bioprocess system. A more powerful elemental analysis is provided by NAA. Unfortunately, this
presently requires a small nuclear reactor with a "rabbit" transport
system. The "rabbit" is a small container, housing the sample,
which is transported through the reactor core by a network of
pneumatic tubing. After a rabbit has traveled through the reactor,
the sample is scrutinized by a scintillation-type detector. As knowledge of nuclear processes become available, we may be able to build
a benchtop neutron generator where the carrier stream of an FIA
would be the rabbit transport and the sample slug would be the
rabbit. This technique would yield an elemental analysis of virtually all of the elements simultaneously.

There is no reason why plug-in complete FIA systems cannot be
small and rack-mountable for easy replacement and repair. The FIA
system of today could have diagnosis based on its own peak-shape
history and response versus standard history. Adaptable and highly reliable FIAs, if modules become inexpensive, could be redundant
and make intelligent decisions using solid-state neural networks.
All of these systems are possible; the preliminary steps have yet to
be taken. In order for FIA to evolve, it needs to be used practically and simply under present day limitations.

REFERENCES

1. Ruzicka, J. and Hansen, E. H., Flow Injection Analysis,
   Wiley, New York (1981).
2. Ruzicka, J. and Hansen, E. H., Flow Injection Analysis, 2nd
   ed., Wiley, New York (1988).
3. Valca'rcel, M. and Lugue de Castro, M. D., Flow Injection
   Analysis Principles and Applications, Halsted Press, Great
   Britain (1987).
4. Hansen, H. E. and Ruzicka, J., United States Patent No.
   4,022,575 (1977).
5. Stewart, K. K., Beecher, G. R., and Hare, P. E., United
   States Patent No. 4,013,413 (1977).
6. Negy, G. Z., Feher, Z., and Pungor, E., Anal. Chim. Acta,
   52, 47 (1970).
7. Ruzicka, J. and Hansen, E. H., Anal. Chim. Acta, 78, 145
   (1975).
8. Stewart, K. K., Beecher, G. R., and Hare, P. E., Anal.
   Biochem., 70, 167 (1976).
9. Recktenwald, A., Kromer, K. H., and Kula, M. R., Enzyme
   Microb. Technol., 7, 146 (1985).
10. Parker, C. P., Gardell, M. G., and DiBiasio, P., Am. Bio-
    technol. Lab., 3(5), 37 (1985).

11. Nikolajsen, K., Nielsen, J., and Villadseen, J., Anal. Chim. Acta, 214, 137 (1988).
12. Recktenwald, A., Kromer, K. H., and Kula, M. R., Enzyme Microb. Technol., 7, 607 (1985).
13. Gornall, A. G., Bardwill, C. J., and David, M. M., J. Biol. Chem., 177, 751 (1949).
14. Henry, R. J., Sobel, C., and Berkman, S., Anal. Chem., 29, 1491 (1957).
15. Bradford, M. M., Anal. Biochem., 72, 248 (1976).
16. Kromer, K. H. and Kula, M. R., Anal. Chim. Acta, 163, 3 (1984).
17. Methods of Analysis, 13th ed., 2.022(a), Assoc. Off. Anal. Chem. (1980).

11

The Use of On-Line Sensors in Bioprocess Control

MICHAEL T. REILLY   *E. I. du Pont de Nemours & Company, Wilmington, Delaware*

MARVIN CHARLES   *Lehigh University, Bethlehem, Pennsylvania*

I. INTRODUCTION

Making chicken soup* or monoclonal antibodies—it's all the same to the extent that all processes share a common characteristic: uncertainty borne of conspiracy between imperfect knowledge and unforeseen circumstances. The best means must be determined by rational evaluation of the cost/benefit ratio given the real objectives of the process.

Bioprocesses are no different. What distinguishes them as a class from other processes are the unique characteristics and requirements of the living organisms they employ. Differences among bioprocesses, and hence differences in the rational control/monitoring systems required, derive primarily from the specific organism(s) or other biocatalysts used, the product(s) manufactured, and the scale and purpose (e.g., lab development versus large-scale production) of the process. In any event, a rational system can be a tremendous asset. For example: at production scale, it can contribute to effecting economically

1. Process and product safety and reliability
2. Conformity with regulatory requirements
3. Overall integration of plant operations

---

*In some circles, a highly developed art form, yielding a product having proven medicinal properties, which product in other circles is manufactured by a process comprising an odd mixture of art and technology conducted in cultlike fashion by practitioners guarding secrets known only to their competition.

4. Obtaining and maintaining process data in useful forms
5. Generating reports required to satisfy regulatory requirements
    and reports useful as bases for process and plant improvements.

This chapter provides enough background to appreciate the basis
for the rational design and implementation of bioprocess monitoring
and control systems. We introduce those characteristics of living
organisms germane to monitoring/control and to basic concepts,
strategies, and hardware of monitoring/control systems. The emphasis, because of scope and space limitations, is confined to fermentation processes and to automated, digital control systems employing some type of digital computer. This is not meant to imply that
such systems are always the rational choice; they are not. In many
instances, other systems, including simple manual control, are currently most appropriate. The fact is, however, that the trend is
toward greater use of automated systems. This trend will yield
considerable benefit so long as automation is well managed and does
not become an end to itself based on too little understanding of
what is being controlled and why.

A. Monitoring/Control Basics (for the Layman)

We might control the process of making chicken soup as follows:

1. Observe (measure) one or more conditions of the process. In
    this case, tasting would give us a measure of taste and temper-
    ature.
2. Compare the observed conditions with what you think they
    should be at any stage of the process.
3. Decide what to do to decrease the difference(s) between the ob-
    servation(s) and the desired condition(s).
4. Act on the decision. In this case, one might add salt and/or
    change the height of the flame.

This procedure is called closed-loop, feedback control. Closed-
loop refers to the action of making the correction based on an observable characteristic of the process (saltiness and/or temperature). Feedback refers to making the correction based on something
that already has happened in the process.
    Now consider a different approach: instead of tasting periodi-
cally, and making the requisite corrections yourself, you instruct
your kid to change the flame height and to add salt in specified
amounts at specific times.
    This approach is called open-loop control. Open-loop control is
riskier than closed-loop, feedback control, requires more chutzpah,

and is premised on the assumption that you know enough about the
process, raw materials, and your kid to be able to predict with
reasonable accuracy what will happen as a function of time. But it
usually is less demanding and less expensive (assuming you consider your time to be more valuable than your kid's time).

The first example also illustrates the concept of explicit control
(also called direct control): you measure directly the variable you
are trying to control. An alternative, indirect (or inferential) control, can be illustrated by the example of cooking a turkey. Here,
we can measure the internal temperature and observe the external
appearance. From these observations we infer the apearance and
the taste of the inside. But we cannot observe directly the internal condition without violating an important constraint: thous shalt
leave the bird intact for the deft hand of the carver.

The chicken soup example also can illustrate the concept of
response time. The pot can, and probably will, boil over (usually
when you are on the phone). How much soup gets out of the pot
depends on

1. How quickly you observe (sense) the boilover or potential boilover. This is the observer or sensor response time. It will
   depend on factors such as visual and/or aural acuity.
2. How long it takes you to decide on the kind and magnitude of
   corrective action to take (e.g., lower the flame, remove the pot
   from the stove), and how long it takes you to initiate the action. Taken together, these actions constitute the controller
   response time.
3. How long it takes for the corrective action to occur: removing
   the pot probably will take the least time (although this action
   may not always be possible without some serious risk). This is
   the response time of the final control element.
4. How long it takes the soup to respond to the control action.
   This is the process response time, and will depend on the size
   of the pot, the liquid level in the pot, and the composition of
   the soup.

The shorter the response times, the less likelihood of subsequent cleanup. But one can go too far. For example, responding
too quickly to potential boilover could have you moving the pot
constantly on and off the range. This sort of oscillatory behavior
probably will not be very good for the soup; it certainly will not be
good for the control system—or anyone related to it.

Finally, we consider the differences between direct digital control (DDC) and distributed control. One can get a sense of DDC
by thinking about supermarket checkout lines. Modern check-outs

employ universal price code (UPC) scanners and automated cash
registers all tied to a central computer which controls everything.
When these systems work, they are very good and provide many advantages to the shopper and to the merchant. But when the computer goes down, and you are standing in line with a cart full of
frozen food, all those advantages tend to become a lot less important when compared to reliability.

Consider another approach: each checkout has its own little
microcomputer containing all the information and function necessary
to operate its own checkout line. It also stores sales records,
which it reports to the central computer on request (for inventory,
bookkeeping, etc.). The central computer transmits to the local
microcomputers updated prices, names of bad check passers, etc.
With this system not much function is lost if the central computer
goes down. This is an example of distributed control. It is usually more expensive to install than DDC and has the appearance of
being more complicated, but it is a lot more reliable. In many
cases, particularly in the control of industrial processes, the benefits resulting from the added reliability more than compensate for
the extra initial capital expenditure.

Figure 1 summarizes schematically, and a bit more technically
than the previous discussion, the steps in an automatic control system. The example shown is for a system including closed-loop feedback and distributed control.

1. A representative sample is taken. This can be a physical sam-
    ple taken through a valve and delivered to an external sensor,
    or the response to an in situ sensor located appropriately (to
    obtain a representative measurement) in the fermenter.
2. The sample is sensed. Usually, this involves the conversion of
    a concentration, temperature, pressure, etc., to an electrical
    voltage or current, usually at a low level.
3. The sensor signal is amplified. This electrical amplification is
    needed (in most cases) because the low-level signal from the
    sensor cannot be transmitted reliably (noise, line losses, etc.)
    to the control system. If at all possible, the amplification
    should be done at the fermenter, and the distance between the
    sensor and the amplifier should be minimized.
4. The signal is conditioned. Conditioning can involve filtering,
    linearizing, converting from an analog signal to a digital signal
    (A/D conversion, required usually because most sensors provide
    an analog signal, and most controllers are digital devices re-
    quiring digital inputs). The signal conditioning may be done in
    the controller unit.
5. In the controller, the measured value of the variable (from the
    sensor/amplifier/conditioner) is compared to the set point (the

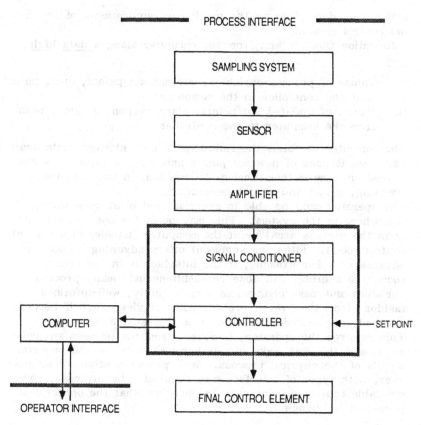

Figure 1 Incorporation of an on-line sensor in a generalized distributed control strategy.

desired value of the process variable). The set point may be entered into the controller manually, or it may be transmitted from the computer. The controller then generates a control signal (usually digital) based on the difference between the measured value and the set point. This signal is used to change a valve position (one example of a <u>final</u> <u>control</u> <u>element</u>) in such a way as to attempt to bring the measured value closer to the set point (e.g., opening a steam valve more to increase a process temperature which is below the set point). The nature and size of the control signal depend on the controller details, which will be discussed later.

6. The control signal output is conditioned. This may involve D/A conversion, voltage level changes, voltage-to-current conver-

sion, etc., depending on the particular requirements of the final control element.

7. Information flows to and from the computer along a <u>data</u> <u>high-way</u>.

    a. Values of process variables, current set points, etc., pass from the controller to the computer.
    b. Values of updated set points, alarm responses, etc., pass from the computer to the controller.

    The computer performs the functions of data storage, data analysis, calculations of new set points and other control variables (based on some mathematical model resident in the computer), trending, graph and report generation.

8. The operator must be able to exercise control at some level, somewhere in the system. This may be at the controller (usually in the process area) or at the computer (usually in a central control room). Either arrangement offers advantages and disadvantages. For example, if the interface is in the control room, the operator can observe simultaneously many process variables and make changes in a more timely, well-informed fashion (from the overall process point of view) than if he or she had only limited information available at one controller. This removes the operator, however, from the process environment, and this may make him or her less aware of the important details of the physical process. With proper configuration, however, both types of interface may be used. In any case, considerable caution is necessary in choosing what the operator is permitted to change.

II. MONITORING AND CONTROL OF FERMENTATIONS

A. Characteristics of Fermentation Processes

Cells grow and form products via a web of interdependent metabolic pathways, each comprising an ordered sequence of reactions, with each reaction catalyzed by one or more enzymes. The product of one reaction can serve as the starting material for other reactions and/or can serve to regulate its own or other pathways.

Pathways are controlled independently and interactively at two basic levels: modulation of the activities of existing enzymes, and modulation of production rates of new enzymes. The characteristic time for activity modulation can be from fractions of a minute to many hours, depending on the organism and on the driving force inducing the control action. The characteristic time for changing enzyme production rates (and hence actual levels) is usually much longer, and can be of the order of the doubling time of the organism. Doubling times range from less than an hour (for many bacteria) to one to several days (for some animal cells). In either case, the control actions are exercised in response to environmental conditions outside the cell and to physiological conditions inside the cell.

These methods of control allow the organism to adapt itself to changes in the environment and in itself. It can adapt not only by changing the rates of individual reactions (and hence some pathways), but it can also "choose" among pathways, to the extent of shutting down or starting up pathways. This ability, unique to living things and without which they probably could not exist, is a double-edged sword for anyone trying to run fermentations productively. It does give an organism enough flexibility to avoid catastrophe (e.g., death) upon environmental changes (which can be effected by the external agencies or by the organism itself). But the response of the organism can vary from an acceptable shift (commercially) in production rate to an unacceptable shift in product types. The nature and extent of the shift will depend on the magnitude and nature of the environmental change, how long it persists, and the inherent characteristics of the organism.

Practical corrective action can be effected if the change can be observed quickly relative to the time scale of the organism and if the mechanics of the system allow the correction to be made in a timely fashion. In most practical cases this corrective action can be done if the change is not very large and if the appropriate sensors are available. One must bear in mind, however, that

1. The rate at which a change can be effected is dependent also on the size of the system: some changes which can be made rapidly in the laboratory (e.g., heating and cooling) are difficult or practically impossible to make rapidly in a large production vessel.
2. Sensors are unavailable for many environmental variables. (This point is very important and will be discussed in greater detail later.)
3. Some changes can be made in only one direction (e.g., one can increase glucose concentration directly but, short of dilution, only the organism can decrease it).
4. Changes important at the level of basic scientific investigation may have little or no significance at a production level, and vice versa.
5. The high degree of homogeneity that can be achieved in small laboratory bioreactors usually cannot be reproduced practically in large production vessels; hence, it often is difficult to obtain

a representative sample or reading. Also, it becomes pointless
to worry about minor variations in temperature, etc., because
there is not much that can be done to improve the situation
without employing impractical methods.

For example, a change in pH of 0.1 unit will not have any practical effects on most production fermentations. Furthermore, this
change can be observed easily, and effective action can be taken
quickly (on the order of one to a few minutes) in either direction.
A similar statement can be made about a temperature change of a
couple of tenths of a degree Centigrade.

Contrarily, the concentration increase resulting from a large
"dump" of phenylacetic acid to a penicillin fermentation cannot be
observed directly because there is no sensor available. Even if
there were, it would not be possible to remove the excess, and the
likelihood is that the organism would not be able to use the acid
quickly enough to allow it to cope with the change. The change
would be irreversible, and the fermentation would be lost.

Effective control requires that the characteristic times of the
physical system (and the control system) be less than those of the
organism. In most cases, the limiting physical characteristics in
fermentations are the practical limits of mixing time and heat and
oxygen transfer rates. Mixing time generally is limited by broth
viscosity, vessel size and geometry, and practical upper limits of
power input. Heat and mass transfer rates are limited by all of
these considerations plus very low driving forces. One must recognize also that very rapid physical response is not always a good
thing unless reliable means are available to prevent harmful over-
shoots. The "dump" of phenylacetic acid is an example.

Medium sterilization is another case where characteristic times
play important roles. The important factors here are heat transfer
rate and the relative rates of microbial death and deleterious chemical reactions (e.g., browning, thermal destruction of vitamins).
In most cases, high temperature increases the rate of microbial
death relative to the rates of deleterious reactions. Using high
temperature allows very short sterilization time which results in
relatively limited effects of the harmful reactions. High heat transfer rates are required to achieve this end, and such high rates
cannot be obtained practically in a fermentation vessel; hence continuous sterilizers employing heat exchangers must be used. But
even here, care and good control must be exercised, since some
harmful reactions are encouraged by high temperature. One example is baking medium onto the walls of the heat exchanger; another
is complex reactions involving phosphate and leading to the formation of products harmful to the fermentation or to precipitation of
other medium components required by the organism.

Generally, fermentations have much longer characteristic times than do chemical reactions. This is one important reason why many of the control strategies and means used so well for controlling chemical reactors are not appropriate for fermenters. Some critical phenomena, however, can occur in a minute or less in some fermentations.

Another major difference between chemical reaction processes and fermentations is the knowledge base relative to the complexity of the system. In general, the ratio is much smaller for fermentations than for chemical processes. Consequently, developing models accurate enough for reliable use in commercial process control is very difficult. This is true even in cases of mature fermentations which have been conducted commercially for many years. In these cases, reliable models tend to be rather empirical. Our rapidly expanding knowledge of organisms is making it possible for us to understand better and to improve these models, but we still are quite far from the point where we can develop reliable predictive models from first principles. Highly empirical models, most likely, will continue to be used for quite some time. The problem, even in the case of empirical models, is exacerbated by the following: (a) the highly nonlinear behavior of fermentations, (b) the batch and/or fed-batch nature of most commercial fermentations, and (c) the highly interactive nature of important fermentation variables.

Concentration is another fermentation characteristic very germane to monitoring and control. The inherent characteristics of living organisms along with practical limitations of heat and mass transfer rates restrict the concentrations (dry-weight basis) of cells, raw materials, and products to 1—50 g/L (up to 100—150 g/L in a few cases). Many components are restricted (because of deleterious effects such as inhibition) to less than 1 g/L, and in some cases to very much less. One example is the requirement to maintain low sugar concentration in the baker's yeast fermentation; high concentrations lead to unacceptably low yields (as a result of alcohol production). Low solubilities limit the concentrations of other important components such as oxygen. Such limitations add considerable obstacles to effective monitoring and control, not the least of which is the development of practical sensors (more later).

Asepsis, required in most commercial fermentations, also makes monitoring and control difficult. Particularly important is that asepsis requires (practically) sensors to be steam sterilizable and to have long-term (12 hr to several days) stability and reliability. These requirements are major obstacles to sensor development.

Among other fermentation process factors which influence monitoring/control system design and implementation are variation in raw materials composition and quality and the host of regulatory requirements which must be satisfied.

B. Fermentation Control

The control of a number of fermentation variables via automatic digital systems has proven to be valuable in many situations. All manner of sophisticated control systems have been proposed, however, for every conceivable fermentation variable. Just because you can control a variable, particularly via sophisticated computerized systems, is no reason to feel compelled to do so. The problem of determining rationally which fermentation variables should be monitored and/or controlled has no pat solution, and must be solved case by case. A few generalizations can be made, but these should be interpreted in the spirit of vague qualifiers. Temperature is monitored and controlled in almost all commercial fermentations; pH is controlled in the vast majority of cases, as are agitator speed, airflow, foam, and back pressure. Dissolved oxygen is monitored but not controlled in most cases. Beyond these variables, all bets are off. Considering the very large number of possible circumstances, it seems most prudent to present here only general factors that one should use in formulating guidelines for specific cases. Among the more important factors are:

1. The objectives of the process and how the system can help to achieve them.
2. The cost and skill levels of the available labor force.
3. The extent to which operators need to have hands-on experience with the process.
4. System reliability. A very important consideration here is the scale and objective (e.g., lab versus plant). In most instances the ratio of reliability to versatility has to be much higher in the plant than it does in the lab. This fact alone (there are others) should lead one to see that equipment suitable for lab use usually is not suitable for plant use. Note also that direct digital control is quite adequate and useful for lab operations, but carries a much higher risk than distributed control at the plant level.
5. The need to keep accurate records that can be validated, particularly to satisfy regulatory requirements (e.g., good manufacturing practice, GMPs). In this same vein, one must also consider the validatability of the entire monitoring/control system, including the software. This point is extremely important.
6. Process and plant scheduling. Surprisingly, this factor is often overlooked. Included here should be a careful analysis of how fermentation control methods will affect the recovery/purification train.
7. All of the factors described previously concerning the nature of fermentation processes. This must include a very careful appraisal of the reliability of process information available.

8. Means by which the system can assist in ongoing process development.

Several examples may serve to illustrate a few of these points.

Baker's Yeast Production

A major factor in yeast production is the sugar concentration in the fermentation broth. Too high a concentration causes the yeast to produce ethanol via the Crabtree effect [1], even under aerobic conditions, resulting in at least two negative effects which decrease profitability. First, sugar converted to alcohol is not available for yeast production (most of the alcohol is stripped out by the air). This decreases yield based on sugar used and hurts process economics considerably because sugar cost is the biggest single item in the total cost of production. Second, any appreciable concentration of ethanol decreases the rate of yeast growth, thereby decreasing fermenter productivity. Historically, sugar concentration has been controlled by use of empirically developed feeding schedules. These methods work, but considerable economic improvements could be realized if sugar concentration could be controlled directly. Reliable, in situ probes are still not available for this purpose. Off-gas analysis methods being investigated measure ethanol concentration in the airstream leaving the fermenter, and this measurement can be correlated quite accurately with ethanol in the broth. The control technique involves feeding sugar on the basis of the off-gas ethanol, and appears to work quite well even though the method is indirect. The instrumentation required is relatively inexpensive, reliable and easy to use. It can be coupled to a computer if desired.

Oxygen Transfer

High dissolved oxygen concentration and/or high oxygen transfer rate required in many aerobic fermentations can be achieved by increasing agitator speed and/or increasing airflow rate. Both usually increase foam generation. The most effective and easiest way to collapse the foam quickly is by addition of a chemical antifoam agent (e.g., silicone, polyethylene glycol). Many of these agents can cause severe fouling of membrane filters often used in the recovery process. Fouling decreases membrane flux, thereby increasing significantly the cost of filtration. Many antifoams (silicones, in particular) can affect adversely cell membranes.

Another approach to increasing oxygen transfer rate (if a suitable antifoam agent cannot be found) is to use higher vessel pressure and/or oxygen enrichment of the incoming air. Both incur additional cost, and oxygen enrichment is further limited by explosion

hazard considerations; nevertheless, use of either or both may turn
out to be the low-cost option when the total production cost (fermentation and recovery) is considered.

Penicillin G Production

Control of glucose concentration is important here because high
concentrations cause severe inhibition of penicillin synthesis [2].
At present, empirical methods are used in conjunction with off-line
measurements to control the concentration because no reliable in
situ probes nor any external/on-line methods have been developed.

Among other important parameters that must be monitored and
controlled is the pH. Good pH control is needed in both the fermentation and the recovery sections. In the latter, penicillin G is
extracted into butyl acetate (for example) from the aqueous broth.
Low pH (less than 2) is required to effect the transfer efficiently
and with a good degree of selectivity [3]. Unfortunately, penicillin
G is rather unstable at low pH; hence, good monitoring and tight
control of pH are essential to achieving good extraction with minimum product degradation.

The general introduction and background discussion is now
complete. We next move on to a more detailed, technical, and
mathematical discussion of control systems.

III. CONTROL CONCEPTS AND CONTROL ALGORITHMS

This section reviews basic control concepts and control algorithms.
Intended is a brief overview of these topics. For a more detailed
discussion of process control, see Refs. 3, 4, and 5.

A. Process Modeling

Prior to devising a control strategy for a process, it is very useful
to construct a simplified process model. Design and implementation
of an effective process control strategy is certainly possible without
a model, but a great deal of insight into the process dynamics results from developing a good quantitative model.

The bases of most models used for process control are the fundamental laws of conservation of mass and energy. These laws may
be approximated by linear, time dependent equations, with the
variable that is to be controlled (the controlled or process variable)
taken as the dependent variable in the equation. The variable
which is adjusted so as to keep the controlled variable at its desired level is the manipulated variable. The process model is rep-

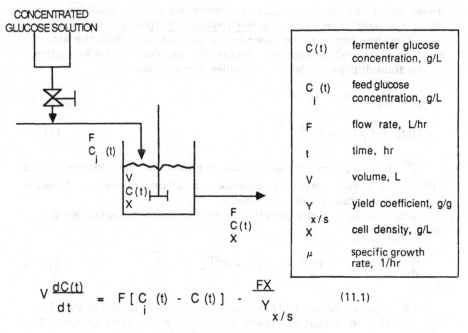

C(t)	fermenter glucose concentration, g/L
C_i(t)	feed glucose concentration, g/L
F	flow rate, L/hr
t	time, hr
V	volume, L
$Y_{x/s}$	yield coefficient, g/g
X	cell density, g/L
μ	specific growth rate, 1/hr

$$V \frac{dC(t)}{dt} = F[C_i(t) - C(t)] - \frac{FX}{Y_{x/s}} \qquad (11.1)$$

Figure 2 Mathematical representation of the glucose mass balance in a continuous fermenter.

resented by a set of ordinary differential equations, with the manipulated variables represented by a set of forcing functions.

Figure 2 shows the development of a process model for a simple example, glucose control in a continuous fermentation. For the purpose of this example, the biomass concentration in the fermenter is assumed to be at steady state, and the yield of cells on glucose is assumed to be constant.

In this example, fermenter volume, inlet and outlet flow rate, cell density, and yield are all constant. The fermenter glucose concentration is the controlled variable, and inlet glucose concentration is the manipulated variable. The desired glucose level in the fermentor is the set point. It may be constant, or it may vary with time; e.g., a high glucose concentration may be required during the growth phase, and a lower concentration during the production phase.

Conventional controller design employs Laplace transforms, and so requires the use of linear equations. Physical processes are generally nonlinear, but often a linear approximation will suffice over a relatively short time interval. Before taking the Laplace transform of the process equation, controlled and manipulated variables are best represented as <u>perturbation variables</u>, that is, as the difference between the time-dependent variables and the setpoint or steady-state values. Taking the example shown in Figure 2, the following perturbation variables are defined:

$$Z(t) = C(t) - C^{sp} \tag{11.2}$$

$$U(t) = C_i(t) - C_i{}^{ss} \tag{11.3}$$

C^{sp} is the glucose concentration set point in the fermenter, and $C_i{}^{ss}$ is the steady-state inlet glucose concentration required to maintain the fermenter at the set point.

Substituting the perturbation variables defined above, we get Eq. (11.4)

$$\frac{dZ}{dt} = D[U(t) + C_i{}^{ss} - Z(t) - C^{sp}] - \frac{\mu X}{Y_{x/s}} \tag{11.4}$$

The dilution rate, D (1/hr), has been substituted for F/V in Eq. (11.4). Note that at steady state, the right-hand side of Eq. (11.1) sums to zero. Therefore, by definition,

$$D[C_i{}^{ss} - C^{sp}] - \frac{\mu X}{Y_{x/s}} = 0 \tag{11.5}$$

Equation (11.4) may now be simplified, as shown:

$$\frac{dZ}{dt} = D[U(t) - Z(t)] \tag{11.6}$$

Laplace-transforming Eq. (11.6) leads to Eq. (11.7), which may be rearranged to yield Eq. (11.8). In Eq. (11.8), $G(s)$ is defined as the <u>process transfer function</u>, which is the ratio of the process output $Z(s)$ to the process input $U(s)$.

$$sZ(s) = D\,U(s) - D\,Z(s) \tag{11.7}$$

$$G(s) = \frac{Z(s)}{U(s)} = \frac{D}{s + D} \tag{11.8}$$

Processes may be represented in block diagram form, with transfer functions being the input/output relationships within the blocks. Figure 2 is a representation of an <u>open-loop</u> process, that

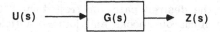

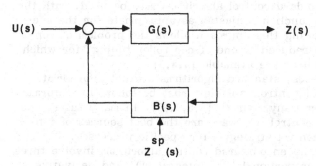

Figure 3 Open- and closed-loop block diagram representations of the continuous fermenter.

is, one where there is no feedback path to complete the loop. If a glucose controller is inserted in the feedback loop to automatically control the glucose concentration in the fermenter, then a closed-loop process results. Both the open-loop and closed-loop block diagrams are shown in Figure 3.

The type of open-loop transfer function represented by Eq. (11.8) is called a first-order lag. The transfer function is a product of the system only, and does not depend on the input type (e.g., step, ramp, etc.). It represents a convenient method of describing the dynamics of the system in compact form. Many other transfer functions are encountered, each with its own characteristic dynamics. Further details are beyond the scope of this chapter; references (4 and 5) contain additional information.

B. Types of Controllers

The glucose controller in Figure 3 is represented by the transfer function B(s). A controller has two inputs: the process variable, and the set point. The controller subtracts the process variable from the set point to generate an error signal. The control algorithm represented by B(s) then acts on the error signal to generate an output, the controller output. This signal then actuates a final control element, which adjusts the inputs to the process as required to maintain it at the set point.

Controller design consists of selecting the algorithm B(s) that causes the process to follow the desired path, or trajectory. Controller design is the topic of numerous books and journal articles and is the subject of a great deal of ongoing research. Good controller design reflects two considerations: the dynamics of the open-loop process to be controlled, and the control objectives. For many processes, standard control algorithms may be used, with the constants for these functions adjusted as appropriate for the specific process. Following this approach allows the problem of controller design to be reduced to one of controller tuning, for which step-by-step procedures are available [4,5,6].

For some processes, standard algorithms are not appropriate, and other methods of controller design must be followed. Information is available from many sources [4,5,7,8]. In many cases, the vendors of process control hardware are the best sources of information concerning control strategies for specific processes.

The most commonly encountered control algorithms involve three modes of control: proportional (P), integral (I), and derivative (D) control action. Frequently, a control algorithm combines several modes, such as proportional-integral (PI) control and proportional-integral-derivative control (PID). Proportional, integral, and derivative are the terms used to describe the method the controller uses to process the error signal to generate a controller output. In the case of proportional control, the controller output is a linear function of the error signal. In other words, the controller output increases in direct proportion to the deviation from the set point. The time-domain and Laplace-domain representations of the proportional controller are given by Eqs. (11.9) and (11.10).

$$M(t) = M^{ss} + K_c E(t) \qquad\qquad (11.9)$$

$$M(s) = M^{ss} + K_c E(s) \qquad\qquad (11.10)$$

M is the controller output, and E is the error signal. Recall that the error is the difference between the set point and the actual process variable. K_c is the controller gain and represents the amount of action that the controller takes in response to a given error. K_c is the parameter which is adjusted when the controller is tuned. The ss superscript indicates the controller output at steady state.

For many processes, proportional control works well. A proportional controller has the decided advantage that tuning and implementation are easy. In many cases, however, proportional-only control is not adequate. A major limitation is that even at steady state, the process variable will differ from the set point by some finite amount. This condition is called <u>steady-state</u> <u>offset</u>, and is a characteristic of proportional-only control. The amount of offset may be minimized by increasing the gain, but doing so results in a more oscillatory process.

Offset may be reduced without increasing excessively the gain by adding some integral action to the controller. With an integral controller, the controller output is proportional to the integral, over time, of the error signal. Equations (11.11) and (11.12) show the time-domain and Laplace-domain representations, respectively, of the integral controller.

$$M(t) = M^{ss} + \frac{1}{t_i} \int E(t) \, dt \qquad (11.11)$$

$$M(s) = M^{ss} + \frac{E(s)}{t_i s} \qquad (11.12)$$

In these equations, t_i is the <u>integral</u> or <u>reset</u> time. It is adjusted when tuning an integral controller. As mentioned, the purpose of integral control is to reduce steady-state offset. The combination of proportional and integral control works well for many processes. It may be shown [9] that PI control is an effective method of controlling substrate in a fed-batch bacterial fermentation.

Integral action does reduce offset, but it does so at the expense of the process dynamics (stability and "smoothness" of the process response to a disturbance). There is no universal solution to this problem, other than to note that controller tuning must balance process performance at steady state (offset) with dynamic response (stability). Compromise is required frequently to satisfy both demands.

Integral action tends to drive the process toward instability. The concepts of stability and instability may be shown to have rigorous mathematical definitions. Stated more simply, however, they relate to the ability of a process to return to an operating point following a perturbation. A stable process will return to its operating point, while an unstable process will not. A good discussion of stability criteria may be found in Refs. 8 and 9.

The third type of control action is derivative control. The purpose of derivative control is to improve dynamic response by calculating controller output based on the time rate of change of the error signal. Equations (11.13) and (11.14) are the time- and Laplace-domain representation of the derivative controller.

$$M(t) = M^{ss} + t_d \frac{dE}{dt} \tag{11.13}$$

$$M(s) = M^{ss} + t_d s E(s) \tag{11.14}$$

The coefficient of the derivative term, t_d, is the derivative time. Tuning the derivative controller involves adjusting t_d. Derivative control action will generally improve process dynamics, except in the presence of excess noise. Taking the derivative of a noisy signal results in generally poorer control.

All three modes of control are frequently combined, resulting in the PID controller, expressed mathematically by Eq. (11.15):

$$M(s) = M^{ss} + K_c \left[1 + \frac{1}{t_i s} + t_d s \right] E(s) \tag{11.15}$$

With a PID controller, the action of the various control modes may be increased or decreased, depending on the choice of the three tuning constants, K_c, t_i, and t_d. The choice of these three parameters may not be arbitrary in all cases, however, because constraints imposed by stability or control objectives will limit the range of values chosen.

C. Adaptive Control

In the ideal case, one would like to select the three PID constants which best satisfy the control objective, and then let the process run forever. This approach may be possible if the process is time-invariant, although no real process is completely so. Many biological processes are notoriously time-varying, particularly when operated in batch or fed-batch modes. To achieve good control in such cases, it may be necessary to retune the controllers during the process. Retuning may be done manually, at discrete intervals, or it may be done automatically, by an adaptive or self-tuning controller.

Adaptive control involves automatic on-line controller tuning, based on a performance algorithm. Adaptive control involves three functions - identification, decision and modification [10]. Identification is the determination, by the controller, of the mathematical description of the process, i.e., its transfer function. Decision involves comparing the process performance to an objective or to a model. Modification involves making some change in the system (the tuning constants) to keep the process at its optimum, as represented by the control objective.

Adaptive control may be particularly appropriate for highly non-linear bioprocesses [11]. Linearization is only appropriate over a relatively short time interval, and so the system must be retuned frequently, as the process moves beyond each linearization interval. Further details regarding adaptive control in general, and adaptive control of nonlinear systems, in particular, are available [10,11,12, 13].

IV. PROCESS CONTROL STRATEGIES

The previous section introduced some of the basic concepts of process models, block diagrams, and control algorithms. This section deals with the methods by which specific control algorithms are implemented into the process. Specifically, it will describe some of the decisions involved in controlling bioprocesses.

A. Open-Loop Control

As discussed earlier, feedback control requires that a measurement of the process variable be available for feeding back information to adjust the manipulated variable. For many bioprocesses, making some measurements is difficult or impossible. In those cases, open-loop or <u>preprogrammed</u> control can be practiced. Here, the manipulated variable is preprogrammed to follow a desired trajectory. An example is the use of a programmed glucose addition schedule for the baker's yeast fermentations.

Open-loop control sometimes works well, when the process is well characterized and relatively invariant. The operator knows from prior experience, for example, how the glucose demand varies with time, and so is able to preset the feed schedule. Open-loop control has the disadvantage that if the process does change, there is no direct feedback information available to allow the inputs to be adjusted. In one example of preprogrammed substrate addition [14], glucose was added to a fed-batch <u>E. coli</u> fermentation based on the assumption of exponentially increasing demand. The required feed rate was calculated based on a fixed growth rate and yield coefficient. The results of this fermentation show that the glucose concentration in the fermenter varied significantly, apparently because of unanticipated changes in growth rate and yield.

Open-loop control is easy to implement, and has been shown to work well in some cases. For highly variable processes, or processes lacking an extensive experience base, feedback control is preferred. When no process measurement is available, however, other options must be explored.

B. Direct and Indirect Control

When the process variable is not measurable, and open-loop control is not feasible, inferential control is sometimes used. Many inferential control strategies, particularly for substrate addition to a fed-batch fermentation, have been reported. Inferential control strategies use measured or calculated process variables to infer the value of another process variable, and then exercise control based on this inferred variable.

An example of inferential control based on a measured variable is glucose control in an E. coli fermentation based on the measurement of dissolved oxygen concentration [15,16]. In this work, the dissolved oxygen level in a fed-batch fermentation was used to control the glucose feed rate. As the glucose concentration became growth-limiting, the growth rate, and hence the oxygen demand, decreased. Consequently, the dissolved oxygen level in the fermenter rose, signaling the need to add glucose. A similar strategy is possible whereby a change in some other nutrient is taken as an indication of glucose consumption. In another E. coli fermentation, ammonium concentration was controlled directly [17], by the use of an ammonium gas electrode. Glucose was then fed to the fermenter at a constant ratio to ammonia.

Another type of inferential control systemizes calculation of the controlled variable based on the use of available measurement data, followed by appropriate control action. Much of the early work in this area involved the strategy of indirect component balancing [18,19,20]. These methods use on-line measurements of oxygen uptake rate (OUR), carbon dioxide evolution rate (CER), and ammonia addition rate to calculate cell growth and glucose uptake based on a material balance and stoichiometric equation. Glucose is then fed as required, based on the model predictions. In some cases, indirect component balancing requires the assumption of fixed yield and kinetic coefficients. In other cases, on-line estimation algorithms are used to arrive at a statistical best fit of these parameters, which then may be adjusted with time.

Indirect component balancing has worked well in certain applications, particularly the baker's yeast fermentations. In general, the approach is best suited to well-characterized systems, for which reasonably good kinetic and stoichiometric models exist. The use of on-line identification methods avoids the limitation imposed by the assumption of fixed yields and kinetics, and improves considerably the method. The disadvantages of indirect component balancing are that a fairly sophisticated computer system is required, and that the method is sometimes difficult to implement.

Control based on the direct measurement of the controlled variable is generally preferred to indirect mehtods. Section VI includes an example illustrating the benefits of the direct control of glucose concentration in a fed-batch E. coli fermentation.

C. Single-Loop and Multiloop Control

The control loop consists of the entire feedback circuit, including the controlled variable, the manipulated variable, and the controller. Many processes consist of several such loops, all operating independently of each other. Such a system is described as having single-loop control. Single-loop control works well when the amount of interaction among loops is small, and when no coordination among loops is required.

Inherent in configuring a multiloop system is the need to determine variable pairing, which is the matching of inputs and outputs for each loop. In some cases, variable pairing is obvious; an example is pH control, where the input (addition of acid/base) is naturally paired with the output (the fermenter pH). In some other cases, various pairing options are available. For example, for temperature control, both the flow rate and the temperature of coolant to the fermenter jacket may be varied. Single-loop control requires that a specific pairing decision be made for each loop.

Several other control arrangements are possible. Two that will be described here are cascade control and multivariable control. They are similar in the sense that some interaction among loops is required, but they address very different considerations.

Cascade control involves two or more controllers, where the output of the first controller serves as the set point to the second controller. An example is the use of a dissolved oxygen controller and an airflow controller in cascade. In this case, the dissolved oxygen measurement serves as the input to the airflow controller. When the fermenter is at steady state, the dissolved oxygen level and the inlet airflow rate will be constant. If the dissolved oxygen level falls below the set point, the dissolved oxygen controller increases the set point to the airflow controller, and the airflow controller sends a signal to a valve on the inlet air line to open further.

It may seem simpler to use the dissolved oxygen measurement as the input to the airflow controller, and hence use just one controller to maintain the dissolved oxygen level directly. Use of a cascade system, however, is recommended for two reasons: First, even when the set point is fixed, the inlet airflow rate may vary, requiring the use of a feedback controller. If, for example, there are fluctuations in the inlet air pressure, the airflow controller must adjust the valve positioning to maintain a fixed flow rate. If the dissolved oxygen signal was used as the input, the airflow controller would not be able to respond to variations in inlet pressure, until those variations were seen as changes in the dissolved oxygen level in the fermenter.

The second reason for cascade control of dissolved oxygen is that there may be more than one control output desired. For example, the dissolved oxygen control strategy may consist of holding inlet airflow rate constant and increasing the agitator speed as needed, up to the maximum speed. Once the maximum agitator speed is reached, then the strategy might require that the airflow rate be increased. This strategy is shown in Figure 4. In this case, the output of the dissolved oxygen controller serves again as a set point, but now to two controllers.

Implementation of the system described above requires ensuring that the low controller output be prevented from reaching the airflow controller, and the high controller output be prevented from reaching the agitator speed controller. High- and low-bandpass filters are used to accomplish these tasks.

A second type of multiloop control is multivariable control. Such control is required when there is a significant amount of interaction among several loops. Mathematically, interaction occurs when the process is described by equations which are coupled through one or more variables. An example is the interaction between substrate and temperature. Consider the case where the substrate is toxic and so must be maintained at a very low level. When this substrate is growth-limiting, a slight change in substrate concentration will result in a significant change in the growth rate, and hence, heat evolution. Consequently, the potential exists for the substrate and temperature control loops to interact and result in oscillations in both substrate concentration and temperature.

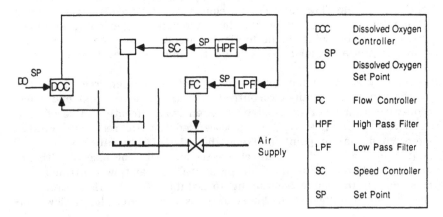

Figure 4   Cascade control of dissolved oxygen.

Other such interactions, for example, among substrate, dissolved oxygen, temperature, and pH may be imagined. In many cases, interactions of this type do not necessarily mean that single-loop control of the variables in question is not feasible. A mathematical procedure may be used to quantify the amount of interaction among control loops [8]. It (as well as empirical evidence) may be used to determine whether single-loop or multivariable control is appropriate.

When there is significant interaction between two loops, several consequences may result. Changing either set point may change both outputs. It also becomes difficult to adequately tune both loops. It may be necessary to sacrifice the performance of one loop to ensure that the other is well tuned. If the amount of interaction is large enough that these consequences are not acceptable, then the design of a multivariable controller may be required. In that case, controller design becomes much more customized, and off-the-shelf algorithms may not be readily applied. The subject of multivariable control is beyond the scope of this chapter. For additional information, see Refs. 8,21,22.

V. CONTROL HIERARCHY

Previous sections dealt with control algorithms and control strategies. Not discussed was the hardware organization for actual implementation of a process control scheme. The present section describes how the controller hardware may be organized, and the next discusses some of the sampling and sensing components that are in use. This discussion is intended to serve as an introduction to the topic. Process control equipment vendors are frequenatly an excellent source of more detailed information.

A. Local Control

The most basic level of control is local control. Local control is generally best suited to single-loop control, although the advent of more powerful, inexpensive microprocessors may expand local control to multiloop systems. Local control involves the use of process controllers located in the actual processing environment. Control is effected by the use of dedicated microprocessor controllers, with one controller serving one or a few loops. These controllers are usually off-the-shelf instruments employing standard control algorithms.

Local control offers the advantage of high reliability. If a controller is dedicated to a single loop, then the failure of that unit results in the loss of control on only one loop. As will be discussed, the failure of highly centralized control systems can result in the loss of control of entire processes.

The disadvantage of local controllers is their limited capabilities. Local controllers generally do not offer any long-term data storage capability and may not be able to interact with one another unless some type of supervisory system is in use. Consequently, one may implement control strategies such as cascade control, but multivariable control is not possible. A further disadvantage is that there is no centralized operator interface. The values of relatively few parameters (set point, process variable, controller output) may be displayed on each controller, but the operator cannot view all parameters from a central location.

Although local control offers the very significant advantage of single-loop reliability, the disadvantages cited above make effective implementation difficult on all but the simplest processes. Therefore, local control is sometimes implemented as but the first component in an overall hierarchy, as will be discussed later.

B. Direct Digital Control

Hardware redundancy is one way to achieve good reliability with local control, but redundancy increases equipment cost. An alternative approach is to eliminate redundant operations by configuring a more centralized system. Direct digital control (DDC) is one method of accomplishing this objective [23,24]. Direct digital control is based on the use of a single processor to control many loops. A digital computer is used, and the control loops are implemented in software. All process variables (generally analog signals) are received by an analog-to-digital (A/D) converter and transmitted over a data highway to the computer. Using algorithms and tuning constants stored in memory, the computer calculates controller outputs based on these inputs. Controller outputs then pass back over a data highway, through a D/A converter, and are relayed (usually) as analog signals to the final control elements.

DDC offers several advantages over local control. Since all loops are controlled by a common processor, many different control algorithms and strategies are possible, including multivariable control. In addition, because tuning constants are stored in software, adaptive control is easily implemented, whereby the outputs of the adaptive control program (the tuning constants) are inputs to the control program. Moreover, the operator is able to view many different parameters from a single workstation. Historical data may also be stored on a peripheral device, and may be viewed, printed, or further manipulated from a central location.

DDC also has several rather significant disadvantages. The major disadvantage is the system's reliability. Unlike the failure of a local controller, the failure of a DDC system results in the loss of control of the entire process for long enough that an entire fermentation batch may be lost—a costly consequence in most cases. DDC was often used with some early computer-coupled fermentations, but now is seldom used by itself for bioprocess control.

As an aside, it should be noted that the distinction between DDC and local control is less relevant today than it was 10 years ago. The key difference between the two is not the use of a computer for DDC versus the use of microprocessors for local control. Rather, the key point is the number of processors in use for control. Having one centralized computer increases the consequence of system failure. Distributing the processing responsibilities among several units increases reliability. Increases in computing power and decreases in hardware cost have made it possible to have many of the advantages of software-based control (adaptive control, data presentation, on-line reconfiguration) while still retaining the reliability of multiple processors. Distributed control systems in use today combine the reliability of local control with the flexibility of centralized control by allowing many local controllers to interact over a common network.

C. Distributed Control

Distributed control systems were developed to combine the reliability of local control with the advantages of centralization associated with DDC. Many variations are possible, but these systems generally use a series of microprocessor-based local controllers, combined with a centralized supervisory system [9,23,24].

The local controllers communicate in two directions over a data highway with the supervisory computer. The computer provides several functions, including data analysis and storage, historical trending and graphics, and report generation. It may also be used to run optimization and adaptive control routines, and the outputs from these routines (new set points and tuning parameters) are then transmitted (downloaded) to the controllers. With this system, failure of an individual controller results in the loss of control to a single loop. Failure of the computer does not result in the loss of control of any loop. At worst, the controllers must default to the last set point (and tuning parameters) received from the computer. Such default, however, is a much more satisfactory situation than the failure of the DDC system, because local control of all loops will continue.

One such distributed control system, in use on a pilot-scale fermentation system, has been reported [9]. In this system, the signals from the sensors and to the final control elements were analog signals. The A/D and D/A conversions were done locally by the microprocessor-based digital controllers. The controllers communicated over a data highway, which in this case employed a standard RS-422 link. Supervision is provided by a DEC PDP 11/73 minicomputer from Digital Equipment Corporation (Maynard, Massachusetts). Communications, adaptive control, data storage, and other routines were written in FORTRAN. Other programming languages may also be used. FORTRAN, however, has historically been the language of choice for scientific and engineering calculations.

This system is shown in Figure 5. The controllers used were local microprocessor-based digital controllers, and were interfaced to the minicomputer using standard hardware and software protocols. Configuring a system such as this one, capable of consistent, reliable operation, requires careful integration of the various system components. Careful integration is needed particularly when hardware from different vendors must be used, and requires attention to software and communications protocols.

A second type of system is available, where both controllers and supervisory system are provided as a package, from a single vendor (cf., 25). Several such systems are marketed, offering many of the same features. The terminology varies with the vendor, but these systems employ multiple control stations with 32 to 64 or more control loops incorporated into one control station. The control stations are linked over a data highway with a central operator workstation. Some data storage may be done at the control station, but trending, graphics, and most other functions are done at the operator workstation.

VI. CONTROL HARDWARE/IMPLEMENTATION

This section will discuss some of the considerations involved in incorporating a control loop or control strategy into a bioprocess. Included is a discussion of sensors, sampling systems, and final control elements. Sensing the variables of interest in bioprocesses involves complications not normally encountered in chemical processes. These differences result from the critical need to maintain process integrity to ensure asepsis and containment. The result of the requirement for asepsis is that potential in situ sensors must be sterilizable, and hence few are available. Containment requires that very careful consideration be given to the disposition of spent sample streams.

A. Sensors

Sensors can be grouped into three broad categories: in situ, external/on-line, and external/off-line. In situ probes are usually

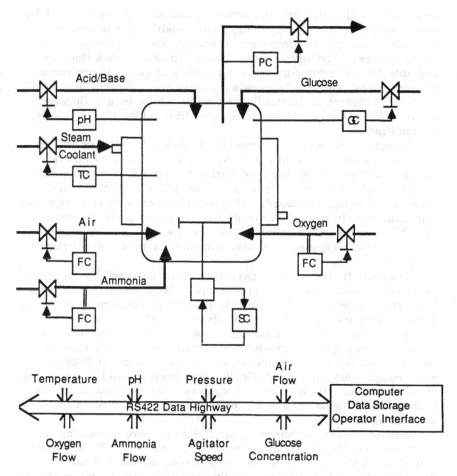

Figure 5 Distributed control system in use on a 30-L pilot-plant fermenter: FC, GC, PC, SC, and TC indicate flow, glucose concentration, pressure, motor speed, and temperature controllers, respectively. (From Ref. 9.)

considered to be the most desirable. The reasons given most frequently are that they (a) give continuous, instantaneous readings (which is particularly important for very rapid fermentations), (b) decrease sample handling, (c) decrease labor requirements, and (d) eliminate sampling, which puts the fermentation at much less risk of contamination. Ideally, these reasons are all valid, but their practical importance varies significantly depending on the fermentation.

In any event, very few in situ sensors are available for practical application (or even for fermentation R&D). The primary reasons for this situation are the problems and constraints imposed by requirements for (a) sterilizability, (b) sensitivity and accuracy at low concentrations, (c) long-term stability, and (d) various regulatory factors. The few reliable, proven in situ probes that are available are primarily for simple physical and chemical environmental parameters, such as temperature, pressure, pH, and dissolved oxygen. Research is proceeding on many others (e.g., fluorescence, FTIR), but it will be quite some time before they are ready for application.

External/on-line sensors provide, in concept, many of the same advantages cited for in situ probes, but do not have to be sterilizable, they can be recalibrated easily during a fermentation and do not require long-term stability. Many other external/on-line systems are being developed. Most employ some form of semipermeable membrane to provide a sterile barrier between the fermenter and the external environment. The systems used employ enzyme (or other electrochemical) probes, and various types of laboratory analytical equipment.

External/off-line sensors have been used successfully for many years for a wide variety of applications. They are quite adequate when the time required for analysis is small in comparison to important fermentation characteristic times and where sampling and sample handling are not significant problems. It is true that (a) there are many gaps in capability and convenience, (b) they do require more labor than would in situ sensors, and (c) they do make record keeping more difficult. But they do meet many of the requirements of automatic control systems. As opposed to in situ probes and external/on-line systems, many proven external/off-line systems are available.

With both external/on-line and off-line sensors, obtaining a representative sample under aseptic conditions is a difficult problem, and is often the major challenge in implementing an automated monitoring system. Sampling will be discussed in more detail in the next section.

There is a considerable body of published, detailed information concerning the current state of sensor technology [26,27,28,29], and a great deal is said about it in this book. A brief discussion here is warranted, however, of sensor technology as it relates to process control. An example of reliable sensors currently available for routine use in fermenters is listed in Table 1.

A brief review of sensors applicable to fermentation and now under development follows:

1. Cell mass probes

   Optical-density-based sensors [25] have been developed, but they usually are not accurate at high cell densities, they do

Table 1 Variables Monitored and Sensors Used in a Typical Pilot-Scale Fermenter

Process variable	Final control element
Temperature	Resistance-type device (RTD)
Pressure	Pressure transducer
Airflow rate	Mass flowmeter
Oxygen flow rate	Mass flowmeter
Ammonia flow rate	Mass flowmeter
Agitator speed	Tachometer
Dissolved oxygen	Galvanic dissolved oxygen probe
pH	pH electrode
Glucose conc.	Glucose (glucose oxidase) enzyme Electrode
Vessel weight	Load cells
Vessel levels	Level probes (microswitches)
Foam Level	Capacitance probe

Source: From Ref. [25].

not distinguish between viable and nonviable cells, and they usually are fouled by various medium components and/or are overgrown by cells.

Flow cytometers are being investigated to monitor individual cellular properties [30,31], but a great deal of development still is needed before they can be considered for anything other than research lab use.

A commercially available in situ fluorescence probe is able to detect reduced pyridine nucleotides (NADH and NADPH) and may serve to indicate cell mass density and metabolic activity [18,32,33]. There is some research evidence [32,33] to show that the fluorescence measurements correlate with cell mass, but the results are not conclusive, and are affected by various environmental factors such as dissolved oxygen concentration. If used in conjunction with other measurements (e.g., oxygen uptake and carbon dioxide evolution rates), fluorescence might be quite useful an indicator of physiological state. Additional work

is needed, however, to demonstrate the reliability and the practical advantages of the probe.

2. Fourier transform infrared spectroscopy (FTIR)

Soluble organics in fermentation broths have been monitored by
FTIR. In one published study [34], ethanol and glucose were
measured by an external/off-line technique, giving results
which differed by 3 to 100% from those obtained by standard
laboratory methods. The technique shows promise of ultimately
becoming a good candidate for in situ monitoring, but needs a
great deal of development. Also, a sterilizable in situ probe
still is not available; hence, automatic, aseptic sampling still is
required.

3. Liquid Chromatography

Soluble organics also have been measured by external/off-line
HPLC [35]. Results were reasonably accurate, but analysis
times were long and automatic/aseptic sampling and injection
were required.

4. Gas Chromatography

Gas chromatography [36] has been used to monitor volatile
substances in samples obtained directly from a fermenter headspace or via a porous polymer (e.g., Teflon^{TM}) tubing probe
immersed in the broth [36,37,38]. The method is relatively
slow and cumbersome, and does not yet have a high degree of
reliability. Similar approaches use mass spectrometry in place
of gas chromatography (GC) and appear to be much more promising.

5. Enzyme Electrodes

Much has been done to develop enzyme electrodes for monitoring
fermentation broths [17,39,40,41,42]. A major limitation, however, is that they cannot be steam sterilized. While a chemically sterilized/aseptically inserted glucose probe (based on glucose oxidase) has been used in situ in a fed-batch E. coli fermentation [9], such use appears to be quite limited. In addition to contamination problems, such glucose probes are sensitive to dissolved oxygen concentration in the broth, which can
cause very inaccurate results. Enzyme electrodes also have
been used in an external/on-line mode, typically in a flow injection analysis (FIA) configuration [39,40,41,43].

6. Affinity/Microelectronics Devices

Hybrid affinity/microelectronic sensors have been used external-
ly/on-line [27,44]. These devices show considerable promise
for reasons of high sensitivities and low cost, based on the use
of microfabrication technology. Currently, only a few analytes
have been measured with such devices, but work is continuing
in this area.

7. Fiber-Optic and Acoustic Sensors

Fiber-optic sensors have been used to monitor protein concentrations and the densities of hybridoma cells in a suspension culture [45,46,47]. Such sensors are not yet available commercially, but their rapid development is being driven by opportunities in medical diagnostics.

Acoustic resonance densitometry (ARD, 48) has been used to measure biomass concentrations based on the dielectric properties of high-conductivity microbial suspensions. The method may be limited to use with solids-free media, but it does appear to show promise.

B. Sampling

Because few in situ sensors are available, sampling becomes a critical, but often overlooked, component of bioprocess monitoring. In some cases, the sampling system is the component that limits the feasibility of on-line process monitoring [49]. This problem is particularly true with tissue culture systems, where considerations of asepsis are most severe.

Several considerations are involved in the design of an effective sampling system. It must be able to deliver the correct sample to the sensor. It must obtain a representative sample from a heterogeneous mixture, and it must not change the characteristics of the sample during transport to the sensor. While these requirements apply whether the sensor is located in situ, on-line, or in an adjacent lab, the focus of the present discussion will be in situ and on-line measurements.

Most samples of interest contain solids suspended in a liquid medium, and frequently a cell-free sample must be delivered to the sensor. In additon, the transport time through the sampling system must be sufficiently small, relative to the response time of the particular system. Consequently, either a high sample flow rate, or the use of very low holdup components, or both, is required.

The use of a high sample flow rate may necessitate that the sample be returned to the fermenter. If this is the case, then the entire fluid path must be maintained in an aseptic state. If the sample is not returned to the fermenter, and if the pressure drop is low enough, then the sample may be obtained by simply operating the fermenter at a slight positive pressure. When the sample is to be returned, a pump is required somewhere in the circuit. The use of a pump introduces an additional degree of complexity which makes reliable and aseptic operation more difficult.

Holdup time is minimized by the use of narrow-bore tubing and low-dead-volume fittings. Both increase the difficulty of delivering reliably a sample over a long period of continued operation, because of the likelihood of plugging. Even air bubbles may result in plugging, as sometimes occurs when a highly aerated medium experiences a slight drop in pressure.

Many different sampling configurations have been reported. The simplest examples involve the use of sensors separated from the bioprocess by a sterile barrier. One example involved the use of a glucose enzyme electrode contained in a sleeve and immersed in the fermenter. The end of the sleeve was constructed of a dialysis membrane [50]. In this case, the fermenter was steam-sterilized with the sleeve, but not the electrode, in place. After the fermenter had cooled, the enzyme electrode was inserted into the sleeve, and then used for on-line monitoring.

Examples of sampling systems where a filter was located in situ have been reported. The porous tubing sensors are one type. Here, porous Teflon^{TM} or other polymeric tubing is located in the fermenter. The pore size of the tubing is sufficiently small to contain the desired organisms in the fermenter and to exclude foreign organisms. A carrier stream flows through the tubing, and the (volatile) analyte of interest diffuses through the tubing and into the carrier stream [36,37,38]. Two other systems both involved filters placed inside the fermenter, and were used to collect samples containing soluble, nonvolatile analytes. The first system employed a dialysis membrane built into one of the fermenter baffles [32], and was used to collect a sample for glucose monitoring via an enzyme electrode. The second used a steam-sterilizable microporous filter located inside the fermenter [9], and was also used for monitoring glucose concentration.

Maintaining a sufficient flux across the filter for an extended period was difficult with all of these systems. Filters of this type operate in a dead-end mode with respect to biomass. Consequently, solids accumulate on the filter medium and reduce fluxes at high cell densities [9]. Automated backwash systems may be required for continued long-term operation.

Improved filter configurations are possible,and a few have been reported. A significant improvement in flux may be obtained by the use of a tangential flow filter [41] to reduce the accumulation of cell mass on the filter surface. Again, considerations of sterilizability and holdup volume limit the choice of available filters.

Once the cell-free sample has been obtained, several modes of analysis are possible. Flow injection analysis is one promising method. With FIA systems, the sample is injected in discrete pulses into a flowing buffer stream and is carried past the sensor. Several commercial systems of this type are available [41,43]. FIA systems are particularly well suited for use with enzyme electrodes. They are able to extend the concentration range over which measurements may be made, reduce the response time of the sensor, and reduce the dependence on coreagents, such as oxygen [9]. One such glucose enzyme electrode/FIA system was capable of operating at concentrations up to 50 g/L of glucose, had a response time of less than 3 min, and was insensitive to variations of dissolved oxygen in the sample over a range of 0—100% of saturation [9].

FIA systems show great promise for fermentation systems. They still require, however, an aseptic link to the process, which usually necessitates the use of some type of filter. Additional work is needed to demonstrate automated, reliable sampling systems, which will operate unattended for extended periods of time. Once such systems have been demonstrated, they may be used in conjunction with many different types of sensors.

C. Interfacing Sensors

Unless a complete monitoring and control system is purchased as a package, interfacing individual sensors to a microprocessor controller or computer monitoring system is necessary. The amount of time and effort spent in assembling such a system is often underestimated, sometimes resulting in considerable delay during startup. To avoid this problem, careful thought must be given to the system configuration, and the types of interactions required among the various components. Once this approach has been taken, it will be easier to determine which system options are appropriate.

For the purposes of this discussion, it is useful to consider the various types of sensor outputs, and the type of interface that each requires. In many cases, a transmitter is required to interface the sensor output to the controller. Transmitters convert a sensor signal (frequency, pressure difference, resistance change) to a control signal, which will frequently be an analog current or voltage signal. The usual signal types are those having 0—5 mV or 4—20 mA output ranges. The transmitter, in some cases, is a separate component, but in other cases is an integral part of the sensor.

Some sensors, such as limit switches and level probes, generate a digital signal corresponding to their on/off state. Both digital and analog signals may be interfaced to controllers or computers via standard input hardware. The same considerations apply to output signal processing as to input signal processing. Output signals are generated by the control system, and are used to manipulate the process. Manipulation is done by changing the position of valves, the speed of motors, and the state of switches. Controller outputs may be either analog or digital signals, and are also generated by the use of standard hardware.

The analog and digital input and output circuit boards, sometimes designated AI, AO, DI and DO, may be located centrally, at a computer, or locally, as part of an intelligent data acquisition device. Such a device is generally a microprocessor which accepts a large number of process inputs and then formats these data for transfer, in blocks, over a data highway. The device also acts in the reverse direction, receiving data from the computer and converting them to analog and digital process outputs.

The data acquisition unit has generally the capability to perform input signal conditioning. Signal conditioning consists of linearization, conversion from electrical units to engineering units, noise filtering, etc. The data acquisition unit often allows other signal processing algorithms to be programmed by the user. Good practice generally involves distributing locally as much of the signal conditioning functions as possible to free the computer to devote its resources to running higher-level routines.

The use of a microprocessor-based data acquisiton system also offers the opportunity for significant savings in wiring costs. All process inputs are hard-wired to the data acquisition module, located in the processing environment, and then only one cable (the data highway) is run between this unit and the remote control room.

Other more sophisticated instruments, such as chromatographs and spectrometers, generate a signal with a higher information content, and so frequently have their own dedicated computer to process their inputs. These intelligent instruments transfer data in much the same fashion as data are transferred from the data acquisition module to the computer, i.e. over a data highway.

Data transfer between processors involves several hardware and software considerations. Each processor must employ the same hardware protocol and must pack the data into blocks in the same manner. Further information regarding the details of the available protocol choices is available in Refs. 51 and 52. Note that several different standards are widely used, and care must be exercised to ensure that different processors are compatible with one another.

In addition to hardware protocol, two processors communicating with one another must use compatible software to send and receive data. In most cases, one of the processors acts as the master system, and the other as the slave. For example, an on-line spectrometer (the slave) may communicate with a supervisory computer (the master). The slave will not continuously send data, but will wait until requested by the master. The master must coordinate each sampling/analysis/data transmission cycle so that data processing and transmission are synchronized. There may be many such slaves on the same data highway, so a protocol needs to be established in order that the slaves may be polled individually.

Many different hardware configurations are possible. Figure 6 illustrates two examples: a distributed control system, and a DDC system. The distributed control system is similar to the system

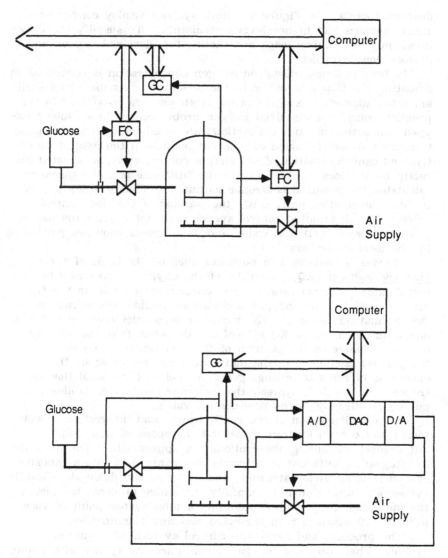

Figure 6 Comparison of distributed control (top) and direct digital
control flow diagrams.

described earlier, in Figure 5. Both systems employ conventional analog sensors and higher-level instruments. To simplify the drawings, only two control loops are shown, for dissolved oxygen and glucose concentration.

In both systems, dissolved oxygen concentration is controlled by adjusting the flow rate of air to the fermenter. As discussed earlier, other dissolved oxygen control strategies are possible. In the present example, a dissolved oxygen probe senses the dissolved oxygen concentration, and the airflow rate is adjusted to maintain the fermenter at the dissolved oxygen set point. In the case of distributed control, position of the airflow control valve is adjusted directly by the local controller. In the DDC example, the computer calculates the required controller output and adjusts directly (via the data acquisition unit, DAQ) the position of the flow control valve. The distributed control system does not require the use of a DAQ, because A/D conversion and data transmission are performed by the local controllers.

Glucose is controlled in both examples on the basis of the res-piratory quotient (RQ), the ratio of the oxygen uptake rate to carbon dioxide evolution rate. A gas chromatograph (GC in the figure) is used to sense oxygen and carbon dioxide concentrations in the exhaust gas stream. The computer then calculates RQ and compares this value to an RQ set point. Deviation from the set point causes a change in the position of the flow control (FC) valve on the glucose feedline. In the distributed control example, this change is effected by changing the set point of the local flow controller. In the DDC system, the computer sends a controller output signal directly to the flow control valve.

Further discussion of the configuration and integration of monitoring and control systems is beyond the scope of this chapter, but several concluding observations are appropriate. The difficulty of integrating different instruments and controllers into a cohesive system is often underestimated, and doing so can delay significantly system startup. Consider carefully the entire process, not simply the individual components. Adopting a total system point of view initially will minimize both hardware cost and startup time.

The process, and hence the control system, will continue to evolve. Therefore, one is wise to configure the system with modular components. This strategy allows for the addition of new sensors and control loops and for the upgrading of controllers and processors. Signal transfer among system components should involve standard protocols and hardware. Doing so greatly facilitates testing and maintenance. A final recommendation is that any new systems be designed to allow for expansion. Especially important is the provision for addition of all types of input and output signals. This provision will become valuable as new bioprocess sensors are developed.

D. Drift and On-Line Calibration

Many fermentations run continuously for several days to several weeks. For these fermentations, drift of the sensor output away from the baseline can become a serious problem. Several approaches are available to correct for drift. The best approach is the use of an on-line recalibration system. Such a system is easiest to implement with a sensor that is located external to the fermenter. Some flow injection systems allow for periodic sensor recalibration [9]. Recalibration is accomplished easily when the system is interfaced to a microprocessor.

In a flow injection system, a microprocessor may be used to sequence the valves used to inject sample and standards into the flowing buffer stream. The standards are injected at a selected frequency, and the microprocessor uses the sensor output following standard injection to generate a new calibration curve. This information is then incorporated into the unit conversion algorithm.

In the case of in situ sensors, on-line recalibration becomes more problematic. Sometimes an aseptic sensor insertion device [9] may be used to withdraw, recalibrate, sterilize, and reinsert a sensor. This method is not practiced widely, however, and is generally perceived as increasing unacceptably the potential for contamination.

Another approach with in situ sensors is to allow the process variable to reach zero and then saturation for a brief period in order to recalibrate the sensor directly. This approach is sometimes possible with dissolved oxygen, but may not be appropriate for other variables.

Some variables are monitored by both on-line and off-line methods. In those cases, the resulting off-line data may be used to recalibrate the sensors. This procedure is somewhat more tedious than direct recalibration, and care must be taken to ensure that the off-line sample represents the conditions inside the fermenter at the time that the sample is analyzed. In the case of sensor drift which follows a well-known trajectory, recalibration may be possible without direct comparison to any standard.

For some sensors, there may be no effective recalibration method, and the only available option is to choose the sensor exhibiting the best long-term stability. For critical variables, the sensors should be tested before selecting the particular type. It is important to conduct these tests in a medium that is as similar as possible to that used in the actual application. For example, some sensors are protected by a membrane which may be fouled by the presence of certain extracellular proteins present in the medium. Fouling increases the response time of the sensor, and may result in the presence of local concentration gradients. If this problem is determined initially by testing the sensor, it may be possible to select a membrane to which protein is less likely to adsorb.

E. Final Control Elements

Control is effected by manipulating chosen variables through final
control elements. The set of final control elements is relatively
small and primarily includes valves, pumps, motors, and heaters.
With this set of elements, all the major variables of interest in a
fermentation may be controlled. Table 2 illustrates how these elements are used to control the nine bioprocess variables listed in
Table 1.

Compared to chemical processes, bioprocesses have the added
complication that the final control elements must usually be sterilizable. Many steam-sterilizable valves are available, but the choice
of pumps is more limited. Pumps offer numerous areas where contaminants may grow and which are difficult to sterilize effectively.
Consequently, they must be selected carefully for the desired application.

In designing the complete control loop, the characteristics of
the final control elements must be considered. Their dynamic response, gain, turndown ratios, and linearity will influence the performance of the control loop. Examples of these considerations,
specifically for control valves, may be found in Refs. 4 and 10.

Table 2   Final Control Elements Used in a Pilot-Plant Fermenter

Process variable	Final control element
Temperature	Valve, steam/cold water lines
Pressure	Valve, exhaust gas line
Airflow rate	Valve, inlet air line
Oxygen flow rate	Valve, inlet oxygen line
Ammonia flow rate	Valve, inlet ammonia line
Agitator speed	Motor, impeller drive
Dissolved oxygen	Valves, motor (air, oxygen, agitator)
pH	Pumps, acid/base lines
Glucose conc.	Pump, glucose supply

Source: From Ref. [25].

VII. EXAMPLE: FEEDBACK GLUCOSE CONTROL

The following example illustrates the use of feedback control of glucose concentration in a fed-batch E. coli fermentation [9]. Data are provided to illustrate the results of controlling glucose automatically, and these data are contrasted with those from a fermentation where the glucose concentration was adjusted manually.

As described previously, substrate control is important with a wide range of organisms, particularly when the fermentation is operated in a fed-batch configuration. In these experiments, glucose was measured with a glucose oxidase enzyme electrode. The electrode was located in an external flow injection apparatus, and was used to measure directly the glucose concentration in a sample stream taken from the fermenter. This system has been designated the glucose flow injection analyzer (GFIA). Accompanying the measuring system was a sampling system, which incorporated a steam-sterilizable microporous filter located in situ.

The enzyme electrode/flow injection system was chosen for several reasons. Enzyme electrodes are simple to operate and relatively inexpensive. The glucose oxidase electrode has been shown to work well in the presence of most of the dissolved organic compounds typically present in a fermentation broth. The major limitation of enzyme electrodes is the thermal lability of the enzyme, which precludes steam sterilization. The glucose electrode is also known to be sensitive to dissolved oxygen concentrations, and its measurement range is limited [9,50]. As will be discussed, however, the use of a flow injection configuration minimizes these limitations.

The overall response time (mixing, transport, sensor) of the measurement system is an important consideration. The response time of the GFIA was about 3 min. The GFIA was coupled to a 30-L stirred-tank fermenter. A low sample flow rate (2 mL/min) was used and obviated the need to return the sample to the fermenter. Consequently, the fluid path through the flow injection analyzer did not need to be maintained in an aseptic state. The components of the GFIA are shown in Figure 7, which illustrates material and information flows and signal processing components.

As Figure 7 indicates, the glucose concentration was measured and then controlled using a microprocessor-based digital controller to adjust the feed rate of glucose to the fermenter. In addition to the glucose control loop, seven other process variables were monitored and controlled. Control was effected by means of a distributed control system. This type of system was chosen for several reasons. First, as discussed earlier, greater reliability results from the use of distributed control, as opposed to DDC. Second, this system offered the greatest flexibility for the present application, because many signal processing functions could be done locally rath-

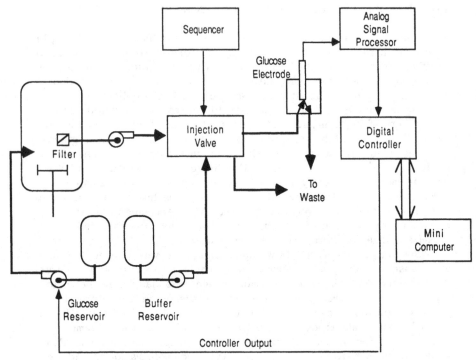

Figure 7 Flow injection analyzer used for feedback control of glucose in a fed-batch <u>E. coli</u> fermentation. (From Ref. 9.)

er than by the host computer. Some of these routines required a much shorter cycle time than was needed for the computer's data acquisition routines. Consequently, the use of the distributed system allowed the different cycle times to be independently adjusted for maximal efficiency. The eight digital controllers communicated with the minicomputer by means of an RS-422 link. Process variables were written to a hard disk every 60 s and were then available for graphics and trending.

Several hardware and software components were required to coordinate the functions of the flow injection analyzer and to generate a glucose concentration signal that could be used by the digital controller. The fermenter was sampled through a small microporous filter located inside the fermenter. The size cutoff of the filter was 0.2 μm, and the total surface area was 225 cm^2. Two methods of controlling the flow rate of sample were tried. First, the fermenter was operated at a slight positive pressure, and a valve was placed on the sample line to adjust the rate of sample flow. The second method involved operating the fermenter at ambient pressure, and using a peristaltic pump to control the sample flow rate through the GFIA. The second method was found to be preferable, because it allowed the flow to be controlled accurately, even when the fermenter back pressure and the flux rate across the filter varied.

An HPLC-type rotary injection valve was used to inject sample into a flowing buffer stream. This valve was controlled by means of a digital sequencer, which also coordinated the actions of the analog signal processor. Because the glucose electrode was exposed to a series of discrete glucose pulses, its output consisted of a series of peaks, with a return to the baseline between the pulses. This raw output signal was conditioned to generate a smooth glucose concentration curve, which was then provided as input to the digital controller.

The analog signal processor served three functions: it provided a constant potential to the glucose electrode, it amplified the low-level signal from the glucose electrode, and it performed a peak-detect-and-hold function to generate the smoothed signal for control. Consequently, this circuit contained three subcomponents: a potentiostat, an amplifier, and a boxcar integrator. The potentiostat provided a constant potential of about 700 mV to the glucose electrode, resulting in an output of about 10 nA. The amplifier converted the current signal to a voltage signal, and generated a 0—2-V input to the boxcar integrator. The integrator received a trigger signal from the sequencer, indicating the beginning of a measurement cycle, and then determined the maximum value of the concentration spike. This value was converted to a 0—10-V output, which was held until the next measurement cycle. The 0—10-V signal was input directly to the digital controller.

The digital controller calculated the controller output required to maintain the glucose concentration at its set point. A 0—10-V output signal actuated an air-operated diaphragm pump on the glucose feedline to adjust the feed rate of glucose as required. An on/off pump was used, with flow regulation achieved by adjusting the width of the on pulses to the pump. The use of this arrangement has several advantages. Good flow regulation was achieved with a relatively inexpensive pump. Second, a very high turndown ratio was possible, which is an important consideration in a fed-batch fermentation, where substrate demand at the end of the fermentation may be 100 to 1000 times greater than at the beginning of the fermentation. Additional details concerning the glucose monitoring and control hardware are available in Refs. 9 and 40. The results of controlling glucose concentration in a fed-batch E. coli fermentation are next described.

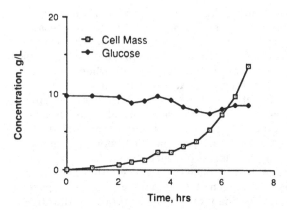

Figure 8 Glucose and cell mass trajectories for a fed-batch E. coli fermentation, with automatic glucose concentration control. (From Ref. 9.)

Figure 8 illustrates the results from a selected E. coli fermentation, where the glucose concentration was controlled with the GFIA system. In this experiment, a glucose set point of 10 g/L was used. Shown in Figure 8 are trajectories corresponding to off-line measurements of glucose concentration and cell dry weight (CDW) concentration. These data illustrate several points. First, reasonably good glucose control was obtained, over a period of about 8 hr. During this time, the cell concentration reached about 25 g/L CDW. There was a deviation from the set point of about 20%, particularly during the latter stages of the fermentation.

The control algorithm used during this fermentation was the proportional-only type. As indicated in the discussion of control algorithms, proportional-only controllers typically produce steady-state offset. The use of PI control would most likely result in improved controller performance, by reducing offset. In these experiments, much of the deviation from the set point was due to fluctuations in the glucose concentration measurement. This portion of the error would not be diminished by better controller design.

The results of the experiment described in Figure 8 may be contrasted with those of a second fermentation, where glucose concentration was controlled manually. Figure 9 illustrates the results of the fermentation run under manual control. In this case, glucose was added as needed, based on off-line measurements, made every 30 min. In both experiments, the same strain of E. coli (K-12) was used, and the inocula were obtained from a common source. All other aspects of the seed preparation, media formula-

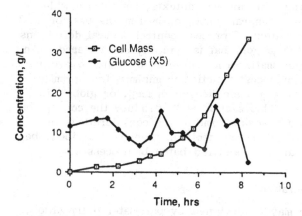

Figure 9 Glucose and cell mass trajectories for a fed-batch E. coli fermentation, with manual glucose concentration control. (From Ref. 9.)

tion, and equipment configuration were the same. The long time lag between measurements resulted in considerably more variation in the fermenter glucose concentration, particularly later in the fermentation, when the substrate demand was high.

In comparing Figures 8 and 9, the benefits of automatic substrate control are difficult to quantify. Despite being run at different glucose concentrations, the specific growth rates and yields of cells on glucose are similar in both cases, indicating a relative insensitivity to bulk glucose concentration. In addition, because these experiments investigated only the growth of cells and not the expression of a product, the overall productivity of the two runs could not be evaluated. Qualitatively, however, it may be observed that the automatic control system resulted in a less variable substrate concentration. For those organisms sensitive to the level of substrate in the bulk, reducing variability in substrate concentration could result in a significant advantage.

VIII. OPTIMIZATION

This section discusses the use of process optimization as a tool to achieve greater productivity in the fermentation process. It will serve as an introduction to the concepts and terminology encountered in this area. The mathematical details of the optimization process may be found in the references.

The term <u>optimization</u>, in various contexts, connotes a wide range of meanings. In a general sense, optimal means best or most desirable [8]. In the context of process control several definitions are needed to clarify the jargon that is typically encountered. An optimization is a rigorous mathematical process by which a profit function is maximized, or a cost function is minimized. Optimizations may be local, involving a single control loop, or global, involving an entire plant. This section will introduce the concept of process optimization and describe some of the considerations involved. The concept of optimal control will then be described, because of its relevance to batch and fed-batch bioprocesses.

A. Single-Loop Optimization

Single loop controllers may be designed by <u>parameter optimization</u>, where tuning parameters are selected on the basis of a mathematical algorithm. In this case, a control structure (e.g., PI) is assumed, and the tuning constants that minimize an error index are calculated. The error is the difference between the set point and the process variable, integrated over time. Many different error indices are used. One example is integral squared error (ISE), which is calculated by taking the square of the error (to convert it to a positive number) and then integrating over time. These types of indices provide information on controller tuning, but do not, in general, provide information regarding the choice of the set point.

After an appropriate error function has been chosen, the tuning parameters which result in minimal error must be determined. Many empirical tuning procedures are available, but parameter optimization is facilitated by the use of a process simulator. A model of the control loop is used to determine the process response to a particular set of inputs. Often, controllers are tuned based on their closed-loop response to a step change in set point or manipulated variable. A given set of tuning parameters, however, is optimal only for a given input. The response to inputs other than a step change may need to be considered.

B. Fermenter Optimization

Controlling an individual fermenter requires controlling several variables, usually 10 to 20. (See, for example, Table 1.) With such a system, independent parameter optimization of each loop will not necessarily result in the optimization of the fermenter as a whole. To perform an overall optimization, it now becomes necessary to first formulate a cost function describing the entire fermenter. Next, the trajectories of all of the major state variables need to be evaluated, and the set which minimizes the cost function of the entire system must be determined.

Because of the number of variables involved, the fermenter optimization process requires the use of a simulator. The available design methods are similar to those used in the design of multivariable controllers. In addition, constraints on the controlled and manipulated variables must be considered. Constraints arise because of limits imposed by process equipment, safety considerations, or basic cellular physiology. The problem may be simplified by using experience to guide the selection of competing strategies and to eliminate relatively insignificant effects from consideration. To this end, a sensitivity analysis may be useful in determining those variables having the greatest effect on the overall process performance.

When the fermenter is operated in batch, optimization may require the consideration of a larger time increment than simply the length of one batch. Scheduling considerations related to the length of the fermentation and the harvesting and inoculation procedure can have a profound effect on overall productivity.

C. Integrated Process Optimization

Optimization of the combined fermentation/recovery process differs from the unit fermenter optimization only in degree of complexity. Many more variables must now be included. In addition, fermentation and recovery processes have historically been operated as separate units, and the interface between these two units may not be easily expressed mathematically. This is particularly true because of modes of operation; fermenters are frequently run in batch or fed-batch modes, while recovery operations may be run continuously. Formulation of the cost function now becomes more difficult; the optimum may require a compromise between fermenter productivity and recovery process yield.

Despite the difficulties, plantwide optimization offers the potential for significant gains. As the processes become more complex, the proper choices from among the various options become increasingly less obvious. Simulation is valuable because it will sometimes show that a seemingly unimportant change in an operating parameter will result in a significant change downstream and a higher product cost. The simulation provides a systematic method of investigating many process options at only a fraction of the cost of pilot plant experimentation. In that sense, process simulation, even without the use of a formal optimization process, provides considerable insight. When coupled to even a simple optimization scheme, however, process simulation becomes an even more valuable tool.

As with the unit fermenter optimization, the scheduling of the fermentation and recovery operations can have a significant impact on productivity and cost, particularly if more than one fermenter is coupled to a given recovery train.

D. Optimal Control

One control method which has received a great deal of attention for use in process optimization is optimal control. Optimal control owes its mathematical basis to Bellman's dynamic programming [53], and Pontryagin's maximum principle [54]. Early applications of optimal control were in the areas of navigation and guidance [8], but it has been applied increasingly to the control of bioprocesses [55,56, 57]. One of the advantages of optimal control is that no a priori assumptions of control structure are required. The problem now becomes one of finding a function which, when applied to the process inputs, will optimize the process performance, relative to a predefined cost function [8]. The details of the method of optimal control may be found in the references, along with information on general features of controller design and process optimization [4,5, 7,8,58] and specific features of bioprocess optimization [55,56,57, 59,60,61,62].

IX. SUMMARY

Commercialization of the products of biotechnology will require a continued emphasis on bioprocess design and scale-up. The use of process automation and control will help to meet the specifications of quality, safety, and cost. In the past, the application of process monitoring and control to biological processes has been limited by the availability of suitable sensors. New developments have combined the tools of microelectronics and molecular biology to address some of these limitations. Continued developments in this area will require close cooperation between microbiologists and process engineers to best apply these new tools. A general understanding of the concepts of process control, as they relate to biological processes, will contribute to that cooperation.

REFERENCES

1. Doelle, H. W. (ed.), Bacterial Metabolism, 2nd ed., Academic Press, New York (1975).
2. Wang, D., Cooney, C., Demain, A. L., Dunnill, P., Humphrey, A. H., and Lilly, M., Fermentation and Enzyme Technology, Wiley, New York (1979).

3. Bailey, J. E. and Ollis, D. F., Biochemical Engineering Fundamentals, 2nd ed., McGraw-Hill, New York (1986).

4. Luyben, W. L., Process Modeling, Simulation, and Control for Chemical Engineers, McGraw-Hill, New York (1973).

5. Coughanowr, D. R. and Koppel, L. B., Process Systems Analysis and Control, McGraw-Hill, New York (1965).

6. Perry, R. H. and Chilton, C. H. (eds.), Chemical Engineers' Handbook, 5th ed., McGraw-Hill, New York (1973).

7. Kuo, B. C., Digital Control Systems, Holt, Rinehart, and Winston, New York (1980).

8. Takahashi, Y., Rabins, M. J., and Auslander, D. M., Control and Dynamic Systems, Addison-Wesley, Reading, MA (1972).

9. Reilly, M. T., "The Development and Implementation of a Feedback Glucose Controller for a Fed-Batch E. coli Fermentation," Ph.D. Thesis, Lehigh University (1987).

10. Forman, E. R., Chemical Engineering Reprint Series, McGraw-Hill, New York (1965).

11. Melin, C., Laanait, A., Delarue, D., and Cordonnier, M., in Modelling and Control of Biotechnical Processes (A. Halme, ed.), Procedings First IFAC Workshop, Helsinki, Finland (1982), pp. 283—290.

12. Chien, I.-L., Seborg, D. E., and Mellichamp, D. A., AIChE J., 33(7), 1079 (1987).

13. Hoopes, H. S., Hawk, W. M., Jr., and Lewis, R. C., ISA Trans., 22(3), 49 (1983).

14. Allen, B. R. and Luli, G. W., A gradient feed process for obtaining high cell densities and control of product expression for recombinant E. coli, in The World Biotech Report 1985, Vol. 2, Proceedings of Biotech '85 USA, Washington, DC (1985), pp. 447—466.

15. Hong, T.-K., Park, S. H., Lee, J. H., Choi, C. Y., and Bae, J. C., J. Ferment. Technol., 62, 49 (1984).

16. Sheehan, J. J., "A Preliminary Evaluation of a Hollow Fiber Microporous Filter for Separation and Concentration of Cell Suspensions," M.S. thesis, Lehigh University (1985).

17. Kole, M. M., Ward, D., and Gerson, D. F., J. Ferment Technol., 64(3), 233 (1986).

18. Zabriskie, D. W., AIChE J., 24(1), 138 (1978).

19. Cooney, C. L., Wang, H. Y., and Wang, D. I. C., Biotechnol. Bioeng., 19, 55 (1977).

20. Stephanopoulos, G. and San, K.-Y., Biotechnol. Bioeng., 26, 1176 (1984).

21. Bollinger, R. E. and Lamb, D. E., IEC Fundamentals, 1(4), 245 (1962).

22. Foss, A. S. and Denn, M. M., AIChE Symp. Ser., 159, 72 (1976).

23. Dartt, S. R., in Practical Process Instrumentation and Control, Vol. II (J. Matley, ed.), McGraw-Hill, New York (1986), pp. 16—19.

24. McMahon, T. K., Chem. Eng., 117 (Oct. 1979).

25. Lee, S. S., and Phillips, J. A., "Application of Distributed Digital Control to a Pilot-Scale Fermentation Unit," American Institute of Chemical Engineers, Annual Meeting, New York (1987).

26. Clarke, D. J. et al., Biosensors, 1, 213 (1984).

27. Turner, A. P. F., Karube, I., and Wilson, G. S., Biosensors, Fundamentals and Applications, Oxford University Press, Oxford, England (1987).

28. Twork, J. V. and Yacynych, A. M., Biotechnol. Prog., 2(5), 67 (1986).

29. Wilson, J. R., Trends Anal. Chem., 3(9), 223 (1984).

30. Phillips, A. P., Martin, K. L., and Capey, A. J., J. Immunol. Methods, 101, 219 (1987).

31. Van Dilla, M. A., Langlois, R. G., Pinkel, D., Yajko, D., and Hadley, W. K., Science, 220, 620 (1983).

32. Zabriskie, D. W., "Real-Time Estimation of Aerobic Batch Fermentation Biomass Concentration by Component Balancing and Culture Fluorescence," Ph.D. Thesis, University of Pennsylvania (1976).

33. Armiger, W. B., Forro, J. F., Montalvo, L. M., Lee, J. F., and Zabriskie, D. W., Chem. Eng. Commun., 45(1—6), 197 (1986).

34. Alberti, J. C., Phillips, J. A., Fink, D. J. and Wacasz, F. M., Biotechnol. Bioeng. Symp., 15, 689 (1986).

35. Dincer, A. K., Kalyanpur, M., Skea, W., Ryan, M., and Kierstead, T., Dev. Ind. Microbiol., 25, 603 (1984).

36. Yamane, T., Matsuda, M., and Sada, E., Biotechnol. Bioeng., 23, 2509 (1981).

37. Kobayashi, T., Yano, T., Mori, H., and Shimizu, S., Biotechnol. Bioeng. Symp. Ser., 9, 73 (1979).

38. Suzuki, T., Mori, H., Yamane, T., and Shimizu, S., Biotechnol. Bioeng., 27, 192 (1985).

39. Chotani, G. and Constantinides, A., Biotechnol. Bioeng., 24, 2743 (1982).

40. Reilly, M. T., Twork, J. V., and Rutstrom, D. J., "The Use of an Electrochemical Sensor for the Control of a Fed-Batch E. coli Fermentation," Federation of Analytical Chemistry and Spectroscopy Societies, 13th Annual Meeting, St. Louis, Sept. 28 (1986).

41. Parker, C. P., Gardell, M. G., and Di Biasio, D., Amer. Biotechnol. Laboratory, 3(5), 37 (1985).

42. Lorbert, S. J., Roettger, B. F., Striebel, J. B., and Ziha, J. E., "An Optimized Glucose Feeding Strategy for Fed Batch Culture of E. coli," American Institute of Chemical Engineers, Annual Meeting, New York (1987).
43. Anonymous, Food Technology, p. 130, Oct. (1986).
44. Janata, J. and Huber, R. J., Solid State Chemical Sensors, Academic Press, New York (1985).
45. Merten, O. W., Palfi, G. E., Staheli, J., and Steiner, J., Dev. Biol. Standard., 66, 357 (1987).
46. Tromberg, B. J., Sepaniak, M. J., Vo-Dinh, T., and Griffin, G. D., Anal. Chem., 59, 1226 (1987).
47. Wolfbeis, O. S., Trends Anal. Chem., 4(7), 184 (1985).
48. Blake-Coleman, B. C., Calder, M. R., Carr, R. J. G., Moody, S. C., and Clarke, D. J., Trends Anal. Chem., 3(9), 229 (1984).
49. Sabelman, E. E., Strategies for Implementation of On-Line Analytical Sensors for Sterile Bioprocess Fluids, ASME Bioprocessing Colloquium, Boston, MA (1987).
50. Cleland, N. and Enfors, S.-O., Anal. Chem., 56, 1880 (1984).
51. Arick, M. R., Data Communications Concepts and Solutions, QED Information Sciences, Wellesley, MA (1987).
52. Halsall, F., Introduction to Data Communications and Computer Networks, Addison-Wesley, Wokingham, England (1985).
53. Bellman, R. E., Dynamic Programming, Princeton University Press, Princeton, NJ (1957).
54. Pontryagin, L. S., Boltyanskii, V. G., Gamkrelidze, R. V., and Mishchenko, E. F., Mathematical Theory of Optimal Processes, Wiley, New York (1962) (Authorized translation from Russian by K. N. Trirogoff).
55. Menawat, A., Mutharasan, R., and Coughanowr, D., AIChE J., 33(5), 776 (1987).
56. Modak, J. M. and Lim, H. C., Biotechnol. Bioeng., 30, 528 (1987).
57. Yousefpour, P. and Williams, D., Biotechnol. Letters, 3(9), 519 (1981).
58. Freeman, E. A. and Abbott, K. M., Proc. IEEE, 116(4), 627 (1969).
59. D'Ans, G., Gottlieb, D. and Kokotovic, P., Automatica, 8, 729 (1972).
60. Guthke, R. and Knorre, W. A., Biotechnol. Bioeng., 23, 2771 (1981).
61. Ohno, H. and Nakanishi, E., Biotechnol. Bioeng., 20, 625 (1978).
62. Staniskis, J. and Levisauskas, D., Biotechnol. Bioeng., 26, 419 (1984).

12

Multicomponent Analysis and Chemometrics for Bioprocess Control

STEVEN D. BROWN   *University of Delaware, Newark, Delaware*

The use of chemometric techniques to aid in interpreting analytical measurements made on bioprocesses has begun very recently. In part, this increase in use may be attributed to the development of a number of new methods of data analysis which are especially suited to treatment of multicomponent data obtained from sensor arrays or from multichannel detectors. Another reason for an increased interest in chemometrics as applied to bioprocesses is the ease with which data can be accessed by the data analysis software now provided with many instruments, and the increased power and mathematical sophistication of that software; all of these changes are made possible with the advent of computers which are inexpensive enough to permit their dedication to analytical instrumentation, but large enough and powerful enough to handle complex data analysis software. Because of the rapid changes occurring in computing power, as well as in the chemometric software relying on that computing power, what seems a complex calculation now may well become much simpler soon. This chapter deals with the goals associated with the application of chemometrics to bioprocesses, and the basics of new chemometric methods under study for use with sensor arrays and multichannel detectors.

I. GOALS OF CHEMOMETRICS IN ANALYTICAL MEASUREMENTS OF BIOPROCESSES

The ideal sensor used for the monitoring of a bioprocess would probably be a completely selective, single-channel device. Multiple sensors could be clustered in a detector module for process monitoring, perhaps sampling the process by some flow-injection scheme. Most sensors are not completely selective, however, and there exists the possibility of a sensor reporting incorrect information because of interferences. It also may be that a completely nonselective sensor is desired, so that unknown products or other species may be detected, and the process control set point altered to change reaction conditions. Whatever the nature of the sensor actually used, it is important to obtain adequate selectivity for the analytes of interest to permit quantitation and control. Thus, an important goal of the use of chemometrics in monitoring of bioprocesses is the enhancement of sensor selectivity.

Even with a completely selective sensor, there exists the need to maintain quantitative calibration. This need is particularly critical when the analyte sensed is important to the control set point. Yet, both systematic and random drift may occur as the sensor changes in its response to the process, or as the composition of the process mixture is altered. Some estimate of the precision of the sensor output is desirable. This estimate might be used in a quality assurance algorithm, as discussed below, or it could be used to assess the reliability of the sensor itself. An estimate of sensor precision depends both on the quality of the original calibration as well as on a description of sensor stability over time. A second goal of the application of chemometrics is the generation of high-quality estimates of sensor precision.

A selective, well-calibrated sensor provides an accurate estimate of some measurable physical property, such as absorbance, current, or potential. The control algorithm may be used to manipulate the process on the basis of other properties, however. The most obvious property is concentration, but product activity (for enzymes) and other, less measurable physical properties may also be important for control. Chemometric methods offer an opportunity for convenient interrelation of the physical properties measured by the sensor and the properties of the bioprocess more relevant to the control algorithm. Insuring the observability of the desired bioprocess properties and the controllability of the process from the sensor measurements is a third goal of chemometrics.

II. ENHANCEMENT OF SENSOR SELECTIVITY

Most sensors are not completely selective to one species; typically, there is a response to two or more species, although the sensitivities toward various species may differ substantially. Compensation for the lack of complete selectivity is often accomplished by use of a multichannel sensor. Either an array of sensors or a more conventional multichannel detector (a diode array or a Fourier transform spectrometer) may be used to provide the additional information necessary to correct for interferences. Chemometric techniques used to quantitate one or more species from data obtained from a multichannel sensor array or detector are classified under the term multicomponent analysis. These multivariate methods of data analysis may be classified into three groups:

1. Methods requiring complete response information
2. Methods requiring only partial component response information
3. Methods requiring no pure component information

Methods in groups 1 and 2 are based on multiple linear regression [1], a well-established statistical technique. They differ primarily in how the model used in the regression step is created, and in what is considered the dependent variable in the regression.

A. Multicomponent Analysis with Complete Model Information

Classical Least-Squares Estimation

The most straightforward approach to the quantitative analysis of several species, by means of their response to an array of J sensors or another type of detector having J sensing channels, is by regressing the multichannel response, S_u, for example an infrared absorbance spectrum of a multicomponent sample, onto a linear model made from the "pure" responses obtained in some calibration step. This multiple linear regression is defined by the

$$S_u = HC_u + E \qquad (12.1)$$

where H is the matrix of the N pure component responses, E is a matrix containing random measurement error, and C_u is the matrix of unknown concentrations (or any property linearly related to the response). A complete list of all symbols, along with matrix/vector dimensions, is given in Table 1. The calibration model H is made by measuring, over the same J channels, a set of N standards, S_c, each consisting of a "pure" individual component present at a known concentration, C_c. Taken together, these concentration vectors form the calibration matrix C_c. Then

$$S_c = HC_c + E \qquad (12.2)$$

The set of equations defined by Eq. (12.2) is solved for H, subject to the usual least-squares criterion that

Table 1   List of Symbols Used

Symbol	Data type and size	Definition
A	N by J matrix	Abstract response matrix
B	v by J matrix	Abstract loading matrix
C_c	N by M matrix	Calibration concentration matrix
C_u	N by 1 vector	Unknown concentration vector
D	L by J matrix	Matrix of sensor responses
E	Matrix (size varies)	Random error matrix
E_c	M by J matrix	Matrix of fit residuals
F(k)	N by N matrix	Kalman state transition matrix
G	N by J matrix	Abstract spectral response
H	N by J matrix	Pure component calibration model matrix
H(k)	N by J matrix	Kalman measurement matrix
I	N by N matrix	Identity matrix
J	Scalar	Number of sensors or sensor channels
K	N by J matrix	K-matrix
K(k)	N by J matrix	Kalman gain matrix
L	Scalar	Number of spectra collected over chromatogram
M	Scalar	Number of standards used in calibration
N	Scalar	Number of estimated components
P	N by J matrix	P-matrix
P(k)	N by N matrix	Kalman covariance matrix
Q(k)	N by N matrix	Kalman state variance
R(k)	N by 1 vector	Kalman measurement noise variance vector
r	N by 1 vector	Target rotation vector
S_c	J by M matrix	Matrix of calibration responses

Table 1 (Continued)

Symbol	Data type and size	Definition
S_u	J by 1 vector	Multicomponent responses
t	L by 1 vector	Target vector
T	M by v matrix	Factor score matrix
U	N by L matrix	Abstract elution profiles
v	Scalar	Number of factors
v	v by 1 vector	PCR and PLS calibration vector
V	L by N matrix	Abstract chromatograms
X(k)	N by 1 vector	Kalman state vector
Z(k)	J by 1 vector	Kalman multicomponent data
ν	J by 1 vector	Kalman innovations

$$\sum_{m=1}^{M} \sum_{n=1}^{N} (S_{mn} - \hat{S}_{mn})^2 = \sum_{m=1}^{M} \sum_{n=1}^{N} E_{mn}^2 \qquad (12.3)$$

is minimized to yield

$$H = S_c C_c^T (C_c C_c^T)^{-1} \qquad (12.4)$$

With the calibration model defined, the set of equations given in Eq. (12.1) is easily solved to obtain

$$C_u = K S_u \qquad (12.5)$$

the classical least squares (CLS) solution [1–6]. In Eq. (12.5) the matrix K, often called the "K-matrix" in the infrared literature [2–4,6], is given by the relation

$$K = (H^T H)^{-1} H^T \qquad (12.6)$$

so that the equation

$$C_u = (H^T H)^{-1} H^T S_u \qquad\qquad (12.7)$$

is used to estimate the desired properties from the observed, unknown responses.

For use of CLS methods to relate a multicomponent response such as a spectrum to the concentrations of the individual components, several restrictions arise, namely that the component responses comprising H must be linearly independent and must span the space defined by S. These mathematical requirements translate to the need to choose sensors or detector channels so that the component responses differ substantially, and to include all possible components—even any background responses—in the matrix H. Typically, all possible components are represented in H by the response of a "pure" component, with solvent responses and other constant responses removed, because mixture of components may lead to multicollinearities in H. While the CLS method works well for the resolution of the effects of multiple, noninteracting species, it fails when unmodeled responses are present, such as unknown background, or when component interactions occur [5,6]. Because CLS methods can use all of a multichannel response, all information can be used in obtaining a set of estimated concentrations. Yet, this feature is not without hazard: fitting of models to multicomponent data can also be done over channels where only noise is present, or where little differences exist between the model components. This "overfitting" of data can produce large errors. To avoid overfitting, channel selection can be done [6,7] so that an optimal model is used to perform multicomponent analysis on data obtained from the most informative sensors.

Using CLS methods it is possible to obtain some information on the nature of poorly modeled responses, however. This feature, as well as the ease with which large response arrays are treated, have made it the method of choice for many multicomponent analyses associated with spectral studies of processes [8,9]. For example, CLS methods have been employed for off-line multicomponent analysis of mid-infrared data obtained on fermentation samples. As indicated in Table 2, the relative errors observed in these analyses are fairly high. One contribution to the error is the variability of the spectral background signal. Spectral backgrounds taken by attenuated total reflectance (the method used to collect data in the table) are very dependent on the nature of the cell surface and the first few micrometers of solution surrounding the sampling crystal. These backgrounds often change over the course of a series of spectral runs as a consequence of source drift, expansion, and contraction of the interferometer because of slight temperature changes, and the slow attack on the sample cell by solvent and solute species.

Table 2 Relative Error from Multicomponent Analysis of Infrared Data on Fermentation Samples

Chemometric method	Spectral range	Relative error in determination of			
		Media	Glucose	Ethanol	Glycerol
CLS pure component model[a]	1160–1000 cm^{-1}	2.9	61.5	12.5	84.8
	2000–1000 cm^{-1}	3.3	43.3	15.1	123.4
CLS mixture model[a]	1160–1000 cm^{-1}	3.2	9.3	29.3	19.4
	2000–1000 cm^{-1}	0.8	22.4	37.0	15.2
CLS mixture model[b]	Local baseline used	2.5	21.6	27.5	27.5
	No baseline correction	22.3	89.0	34.9	58.9
ILS[c]	1160–1000 cm^{-1}	2.1	19.5	18.9	13.8

[a]CLS analysis of selected data collected with 2 cm^{-1} spectral resolution.
[b]CLS analysis of full spectrum, using 6 cm^{-1} spectral resolution.
[c]ILS analysis of Fourier-transformed data over 1160–1000 cm^{-1} spectral range.
Source: Adapted from data in Ref. 8, by permission.

Often biological materials adsorb on the sampling crystal, leading to substantial changes in the effective background [8,9]. Because these background changes are not easily predicted, they are not easily modeled, and CLS methods do not perform well unless some means is provided for continual updating of the effective spectral background. Another contribution to error in CLS estimation results from the interaction of many of the component species. If unexpected, these interactions lead to errors in estimation of analyte concentrations because they are not part of the model, and even if expected, they can be the cause of deviations from the linear (Beer's law) calibration unless proper corrections are made. If interactions are expected, however, they can easily be made part of the model by calibrating the model with mixtures instead of pure components. In that way, the component interactions are built into the multicomponent model, and Beer's law is corrected for the interaction effects. The data in Table 2 show the effect of a strong interaction between the major components (growth media and ethanol) and the minor components (glucose and glycerol). Estimating minor species' concentrations by using CLS with a pure-component model leads to large errors in the concentrations of the minor species, since their spectra are most altered from those used for calibration. The same data, when analyzed using CLS with a mixture model, lead to estimates with substantially lower errors in the estimates of glucose and glycerol because those interactions are better modeled. However, use of mixture-based models worsens estimates of the major components, probably because the calibration mixtures contained amounts of minor species in excess of that found during the multicomponent analyses. In this case, the pure component spectra better represent the responses of major components contained in the mixtures subjected to analysis, and a mixture model would be expected to perform less well than the simpler pure-component model. This example underlines the conflicting criteria which must be considered in using the K-matrix approach.

CLS methods are especially attractive because they are widely available. Several sources exist, but the most direct is to use a good statistical package, either on a mainframe, a minicomputer, or a microcomputer. Most statistical packages offer multiple linear regression methods suited to use with large data sets. Examples include SPSS, SAS, SYSTAT, and STATA. Those familiar with programming can also easily adapt the well-known weighted least-squares algorithm found in any introductory statistics or chemometrics text. Most vendors of infrared instrumentation sell versions of the CLS algorithm as curve-fit programs or as programs for multicomponent analysis. Whatever the source of the program, using it is simple. It is only necessary to specify the model by providing the calibration data, and then to apply the CLS model to the analytical data.

Inverse Least Squares

A second approach to the problem of extracting concentrations from a set of sensor responses is based on an alternative form of the linear relation between the sensor response and the property of interest. In this equation, the concentration is regressed onto a model consisting of the unknown mixture responses, weighted by a calibration matrix. Because of this "inverse" relationship, the method is called inverse least squares (ILS). Like classical least squares, the solution is obtained by multiple linear regression, and sensor calibration precedes the estimation of components in the unknown. The set of calibration responses are expressed as

$$PS_c + E = C_c \tag{12.8}$$

where P, often called the "P-matrix" in the infrared literature, contains information on the "pure" calibration components [10—14]. Subject to the usual least-squares criterion, the solution is

$$P = C_c S_c^T (S_c S_c^T)^{-1} \tag{12.9}$$

Since, for the unknown responses

$$PS_u + E = C_u \tag{12.10}$$

the unknown properties can be directly estimated by

$$C_u = PS_u = C_c S_c^T (S_c S_c^T)^{-1} \tag{12.11}$$

and the pure-component responses contributing to the unknown response can be estimated from the relation

$$H^T = S_c (C_c^T C_c)^{-1} C_c^T \tag{12.12}$$

Unlike the CLS methods discussed above, ILS methods can be used to obtain quantitative information on well-modeled components even in the presence of unmodeled responses, such as background, because concentration vectors may be obtained one at a time from Eq. (12.10). This relative insensitivity to background has been reported to be especially advantageous in FTIR analyses of aqueous biological samples, where ILS methods performed better than CLS methods [9]. ILS methods cannot be used to analyze all data present in large arrays of responses, such as spectra, however, because the number of calibration responses must exceed the number of channels used to sense. Thus, useful "spectra" cannot be obtained though direct use of Eq. (12.12). The nature of any unmodeled response is equally inaccessible. One approach that has been used to circumvent this limitation involves application of ILS methods on Fourier-transformed data. Usually, relatively few Fourier components are needed to describe a multichannel response like an infrared spectrum, and inverse regression can be done on those few Fourier points without appreciable loss of the information present in the full response [12–14]. In fact, Fourier-transformed data were used to obtain the results in Table 2. The limited number of channels used in an analysis based on inverse regression imposes other demands on the user. The bands must be carefully chosen to avoid overfitting (modeling of noise and errors in the calibration data) and to optimize the information available on the components present in the multicomponent model. Further, to insure that the calibration response matrix is rank J, each of the J rows of M calibration responses must be linearly independent, forcing the analyst to ensure that no pair of sensor channels have responses that covary. With spectral data, this requirement is not too demanding, but it restricts the use of ILS methods in reduction of data from other sensor types where data are more correlated [15].

Inverse regression methods are not as widely available as CLS methods, in part because inverse regression is a relatively new concept. While most of the commercial statistical packages do not offer a preprogrammed route to ILS methods, some can be adapted to create the ILS model, which can then be applied to the analytical data using standard methods. It is also easy for those who program to develop an ILS method, either from the equations given here or from the original literature. Although the method is not as well publicized as CLS methods, those capable of coding matrix operations—either directly or by use of packages such as the IMSL library—should have little trouble. Some vendors of infrared instruments also offer ILS software for those seeking preprogrammed versions. However, given the difficulties associated with ILS methods, the author suggests use of an alternative method, such as one of those discussed below.

Principal Components Regression and Partial Least-Squares Methods

Two alternative approaches to multicomponent resolution use factor calibration rather than mixed-component or pure-component responses. Because both can be implemented with versions of the same algorithm, they are discussed together [5,16]. In these methods, the factor calibration step is accomplished by an eigenanalysis. An "inverse" multiple linear regression step follows. In both of these methods, the calibration spectra are represented by the equation

$$S_c^T = TB + E_c \qquad (12.13)$$

where B is a matrix, with rows consisting of "factor loading vectors" obtained from factor analysis of the covariance matrix $S_c^T S_c$. The matrix T contains "factor scores" (intensities) in the coordinate system defined by B. The matrix E_c contains the residuals of the fit of S_c by TB. The matrix of intensities in the abstract factor space is related to concentration with a separate inverse least squares analysis like that in Eq. (12.10). To avoid the need to calculate the quantity $(S_c^T S_c)^{-1}$, the following is solved by least squares

$$Tv + E = C_u \qquad (12.14)$$

where v is a vector relating the factor scores to concentrations. That relation is established by the calibration step, since for each of the calibration vectors comprising C_c

$$v = (T^T T)^{-1} T^T C_c^T \qquad (12.15)$$

Because the factor score matrix is orthogonal, the inversion in Eq. (12.15) is trivial ($T^T T$ is diagonal), and the problems with multicollinearity encountered in ILS are avoided, so all useful response information can be employed. Like CLS and ILS, these methods express an set of unknown, multicomponent responses in terms of two smaller matrices; unlike the more conventional linear regression methods, however, these matrices are not directly related to the "pure" calibration spectra. Thus, these two methods can be considered hybrid methods, combining the calibration advantages of ILS methods with the treatment of the full response offered by CLS methods [5,14,16—17].

One of these methods is called <u>principal components regression</u> (PCR). In PCR, the rows of B are the eigenvectors of $S_c^T S_c$; here, B is an orthonormal matrix. If all eigenvectors of $S_c^T S_c$ are used in B, the results of a multicomponent analysis will be identical to those obtained from multiple linear regression using the P-matrix method. However, if fewer eigenvectors are used—those corresponding to signal and not noise—better multicomponent analysis can be obtained, since the data compression accomplished by dropping "unimportant" factors effectively enhances the quality of the calibration responses used as models for the regression step. In PCR, the reduction of noise is optimal in the least-squares sense, but the loading vectors are not directly related to any single component. This reduction in response error makes it possible to satisfy an important assumption made in ILS methods: namely, that the error in concentration is dominant. The loading vectors cannot provide information on the pure components, but because T is orthogonal, the data analysis can be performed one component at a time without the possibility of encountering problems from multicollinearity [5,16—18].

The other method is partial least squares (PLS). Like principal components analysis, the matrices T and B are generated from a factor analysis of $S_c{}^T S_c$. However, in PLS the concentration information associated with the calibration standards is also used to build the model used in the inverse regression step. This results in an orthogonal matrix T and a nonorthonormal matrix B; these matrices are, in general, different from those produced by PCR because they contain calibration information on concentrations (or similar properties) as well as information on responses. With these matrices, the reduction in response noise is not necessarily optimal, but now the loading vectors in B relate to the components of the multicomponent mixture. By using PLS, loading vectors are generated which are better representative of the pure components of the mixture. Like PCR, PLS offers efficient noise reduction through these loading vectors, allowing better estimation from the inverse regression step indicated in Eq. (12.10). In PCR, however, the loading vectors may be better at representing response, rather than concentration, information. In principle, this modeling method provides PLS with an advantage over other methods for estimating concentrations from multicomponent data; in practice, that advantage is seldom realized [14,17—21]. PLS methods require careful (statistical) design of calibration data sets to obtain optimal performance. When data are limited, PLS methods may be no better than the simpler, faster CLS methods. In most detailed studies comparing PLS and PCR, there has been little difference in the accuracy of the quantitative results [17—19, 21]. PLS generally requires fewer factors than PCR to describe the factor model, however, and the simpler model provides a considerable savings in the time required to calculate the calibration model. This time savings can be important because both PLS and PCR are computationally intensive methods, taking considerable time to converge on a microcomputer.

To date, there has been no direct comparison of PCR, PLS, and CLS on data from bioprocess analysis and bioprocess control. A comparison is available on spectroscopic data, however [19]. Results obtained in that comparison are summarized in Table 3. The PLS and PCR models were constructed as discussed above, using a cross-validation approach to determine whether the addition of a factor improved the predictive ability of the calibration model. In cross-validation, one calibration sample is omitted from the set used to build the model, and the resulting model is used to predict that sample's concentration. The mean square error of prediction determines the correct number of factors for the model. From the table, it is apparent that PLS and PCR methods produced significantly better results than CLS methods for analysis of B_2O_3 and P_2O_5 in borophosphosilicate films. The spectra contained strong interactions between species, as well as large variations in background, making CLS analysis fairly poor. In contrast, the PCR and PLS results show good precision. It should be noted, however, that as many as 10 factors were needed to model this three component system (SiO_2 was the third component), indicating that many factors were needed to describe the interactions and the background variations. The PLS model required fewer factors than the PCR model, but to 0.25 confidence, the results from the two methods were not different.

The results shown above suggest that PLS and PCR methods offer a significant advantage over other regression-based methods. Although PLS methods are very recently developed, they should be considered for data obtained from poorly modeled systems. The algorithms for PLS and PCR are relatively straightforward, but the model-building step is involved, making development of PLS and PCR software a task for experts. Fortunately, several groups involved in development of PLS software offer commercial programs. One, UNSCRAMBLER, was developed by Martens for use with near infrared instrumentation; it has now been generalized for other uses [22]. The venerable package ARTHUR, developed by Kowalski's group, now includes a module for PLS and PCR analysis [23]. Many vendors of spectroscopic instrumentation also offer versions of PLS, although theirs tend to be rather spartan versions.

Methods Using the Kalman Filter

Methods based on classical or inverse regression require the inversion of a matrix to relate the set of unknown, multicomponent responses to concentrations. This matrix inversion step can be a major source of difficulties, ranging from the introduction of computational error to the failure of the analysis because of the inability to invert a singular (or nearly singular) matrix resulting from multicollinearities in standard responses (CLS) or concentrations (ILS). The inversion step is also demanding of memory and compu-

Table 3 Errors of Prediction for Calibration Data from Multicomponent Analysis of Borophosphatosilicate Films

		Squared error of prediction	
Chemometric method	Data pretreatment	B (% w/w)	P (% w/w)
CLS pure-component model[a]	Thickness scaled[b]	0.14	0.60
PLS model[c]	Thickness scaled[b]	0.12(7)	0.27(13)
PLS model[c]	None	0.11(8)	0.22(9)
PCR model[c]	Thickness scaled[b]	0.12(10)	0.27(15)
PCR model[c]	None	0.11(12)	0.21(14)

[a]Using linear approximation of baseline from 425 to 1600 cm^{-1}.
[b]The spectrum of the Si support was removed, then spectral data
were scaled for thickness.
[c]The number in parentheses designates the number of factors required for the calibration model.
Source: Adapted from data in Ref. 19, by permission.

tation time, especially for large matrices resulting from many calibration samples and many-component mixtures. An alternative
method uses a more empirical means of estimating concentrations
from multicomponent responses. As before, multiple linear regression is used. In this case, though, the unknown multicomponent
response is regressed onto the calibration responses, each weighted
by a quantity to be determined. Thus

$$S_u = S_c X + E \tag{12.16}$$

where X is the weight factor. A direct, least-squares solution for
X gives

$$X = (S_c^T S_c)^{-1} S_c^T S_u \tag{12.17}$$

so that the unknown concentration is

$$C_u = C_c X = C_c (S_c^T S_c)^{-1} S_c^T S_u \tag{12.18}$$

Now, the matrix inverted is N × N, the same as in the ILS method. However, in this case the data are overdetermined in sensors, not in calibration spectra, and it is simpler to avoid problems due to multicollinearities.

The more empirical method affords another, more significant advantage: for many data sets, it can be done recursively, in real time, because the calculations can be made simple and fast. Unlike the more statistically based modeling methods mentioned above, this more empirical, recursive approach permits estimation of fluctuating concentrations, background, and nonstationary noise in the response; this method also permits changing models as data are processed to account for sensor changes and nonrandom changes in component concentrations in the mixture, even those altered by a control action [24].

The formalism used to accomplish this empirical modeling is based on that used for state estimation in engineering. In this application, the term "state" refers to a scalar or vector containing the properties to be estimated from the multicomponent response. The method used to estimate the state is based on the Kalman filter, a recursive digital filtering algorithm [24,25]. The Kalman algorithm is summarized in Table 4. In addition to the recursive action of the filter, there are several notable features, including the two equations used to represent the model for the behavior of the state (Eqs. (12.19) and (12.20)), the possibility of noise in the model describing state behavior, as well as the need for an "initial guess" for states and their covariances to begin the recursive estimation process. Other, less obvious features of the filter algorithm are the possibility of treating the measured quantities as either a sequence of scalars, or as a vector, depending on how correlated the measurements are to one another. When the measurements are not highly correlated, a scalar algorithm may be used to process the data without the need to perform matrix inversion [26]. When measured data are too highly correlated to allow treatment as a packet of scalar data, high-precision versions of the Kalman filter, such as the Potter-Schmidt or the Bierman-Thornton algorithms [26, 27], may be used to perform the calculations.

Kalman filter methodology is similar to that used for CLS and ILS. Some of the larger mathematics packages (IMSL, for example) provide the filter algorithm for those wishing to program their own filter. Many references in the engineering literature describe the programming of Kalman filters as well [27]. Other than specialized control packages, no commercial, preprogrammed Kalman filters are available for chemical applications, however.

Table 4 The Kalman Filter Algorithm

Kalman filter model

$$X(k) = X(k - 1) + \Gamma(k - 1)w(k - 1) \quad \text{System} \qquad (12.19)$$

$$Z(k) = H^T(k)X(k) + v(k) \quad \text{Measurement} \qquad (12.20)$$

State estimate extrapolation

$$X(k \mid k - 1) = F(k - 1)X(k - 1 \mid k - 1) + \Gamma(k)w(k) \qquad (12.21)$$

Error covariance extrapolation

$$P(k \mid k - 1) = F(k - 1)P(k - 1 \mid k - 1)F(k - 1)^T$$
$$+ \Gamma(k - 1)Q(k - 1)\Gamma^T(k - 1) \qquad (12.22)$$

State estimate update

$$X(k \mid k) = X(k \mid k - 1) + K(k)(Z(k) - H^T(k)X(k \mid k - 1))$$
$$(12.23)$$

Error covariance update

$$P(k \mid k) = (I - K(k)H(k))P(k \mid k - 1)(I - K(k)H(k))^T$$
$$+ K(k)R(k)K^T(k) \qquad (12.24)$$

Kalman gain calculation

$$K(k) = P(k \mid k - 1)H^T(k)(H(k)P(k \mid k - 1)H^T(k) + R(k))^{-1}$$
$$(12.25)$$

B. Multicomponent Analysis with Incomplete Models

Many samples respond at a sensor array differently than expected from the calibration step, either because of the presence of uncalibrated (and probably unknown) components, a change in the sensors' "background" response, or both. Chemometric methods which are robust to these situations are an important aspect of multicomponent analysis. Because the model is built on the assumption that all components are known and unchanging, the simple CLS method is prone to error when unmodeled responses, of any sort, are present. However, there are extensions to the simple CLS algorithm presented here: it is possible to simultaneously fit the baseline with a separate least-squares model, to include a nonzero intercept in the linear model, and to variance-weight results from all response channels so that channels with poorly modeled responses are less influential on the final estimates. Using this method, CLS analysis fails only when all channels contain unknown interferences, and all interferences cannot be modeled with the nonzero intercept in a linear response versus concentration model, nor fit by the baseline model [4,15]. Stepwise multiple linear regression, using models measured over channels where interference is minimal, can also be used to avoid errors due to incomplete models.

The ILS method enables the more direct compensation of model errors caused by unexpected responses, but this method requires that the number of sensor channels used must match the number of calibration samples prepared. If there happens to be an unmodeled response at one of the preselected channels, its correction is possible, because component concentrations are extracted one at a time; however, since the number of channels is limited by the need to make up and run an equally large number of linearly independent calibration mixtures, there is the likelihood that any unexpected component will go undetected by the few sensor channels used. The use of Fourier-transformed data may help in reducing the restrictions imposed by the nature of the ILS regression step [14].

When the modeled response is constant, but an unmodeled response is also present, PCR and PLS methods can be used to remove the response and to perform multicomponent analysis of the well-modeled components, again because component concentrations are extracted one at a time. Unlike the ILS method, PCR and PLS methods allow use of the full sensor array response to calculate the model during the calibration step so that unexpected responses may not go undetected. Experience with these methods suggests that fairly effective correction of background and other unmodeled components is accomplished. In some cases, qualitative information regarding the nature of the unmodeled responses can be obtained from PLS methods [17–20].

Unexpected responses can also be treated with Kalman filter methods. Since in these approaches, the component concentrations are not extracted individually, the calibration model must be corrected for the presence of the unanticipated response. This correction is accomplished through the use of an adaptive filter [24,28, 29]. In an adaptive filter, the weighting of data is determined by how well that data can be predicted by the filter model; that is, correctly modeled data have a strong influence on state estimates, while poorly modeled data have little or no influence. The weighting is accomplished by examination of the filter innovations sequence, as defined by the relation

$$v(k) = H^T(k)X(k \mid k - 1) - Z(k) \qquad (12.26)$$

when the model is correct, the sequence v should resemble "white noise," but when the model is in error, the sequence will show a systematic component. This method requires a response which has at least one channel where no model error is present, a requirement made more tolerable with the use of derivative and other transformations of data prior to filtering. Correction of models occurs by adapting the measurement noise variance $R(k)$ in Eq. (12.25); when model error is low, $R(k)$ is determined by the actual measurement noise, but when model error is present, $R(k)$ is increased in proportion to the size of the model error, effectively deweighting the poorly modeled data in the recursive fitting process. Figure 1 shows derivative infrared spectra of glucose in ethanol-water solvent. Only the spectrum of glucose, taken in a water solvent, is available from the calibration step [30]. The derivative infrared spectrum in A is fit with a conventional Kalman filter, using the glucose calibration model, leading to 20% error in the estimated glucose concentration. Since the ethanol and water are not included in the model, their concentrations cannot be estimated. In B, the same derivative spectrum is fit with an adaptive filter, using the same incorrect model, but now the error in the estimated glucose is 2%. The use of an adaptive filter permits approximate modeling in multicomponent analysis and avoids some of the restrictions imposed by regression-based modeling. Much the same approach can be applied to other instrumental responses if an array of sensors is used. For example, it is possible to extract the response of a well-modeled component from an unknown (and poorly modeled) mixture measured by diode array spectrometry. This configuration is especially convenient for tracking transient responses, such as those found in liquid chromatographic effluent. In this way, the use of an adaptive filter permits quantitation of components measured in unoptimized separations, a great time saver. It also per-

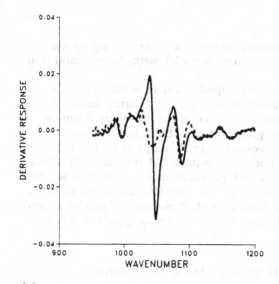

(a)

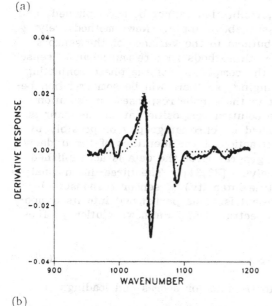

(b)

Figure 1 Fitting of first-derivative infrared data with incomplete models: (a) Glucose in ethanol-water solution (solid line) shown fitted with an ordinary Kalman filter (dashed line). A calibration model containing only the glucose spectrum was used. (b) Glucose in ethanol-water solution (solid line) shown fitted with an adaptive Kalman filter (dashed line). A calibration model containing only the glucose spectrum was used. The adapted component, representing unmodeled species' contributions to the observed spectrum, is also shown (dotted line).

mits tracking of species concentrations in a time-varying system where statistical calibration required for PLS methods is impractical or impossible [31].

Fuzzy modeling may also have application to multicomponent analysis with imperfectly known models [32]. A fuzzy model is one whose concentration - response relation is not given by a single constant (the slope), but is given by a distribution of values. Fuzzy regression, a form of robust regression, can be used to relate responses to concentrations. Classical and inverse calibration methods can be used with fuzzy regression, but no detailed studies of fuzzy modeling have appeared in the chemical literature to date. No commercial software is available, but those comfortable with programming should be able to develop fuzzy modeling software from the reference cited [32].

C. Multicomponent Analysis When No Model Is Available

When no model is available, calibration cannot be accomplished, and none of the methods discussed above apply. Now, methods relying on the extraction of contributions to the variance of the sensors' responses must be used. Such methods require multichannel sensor data over some time where the composition of the phase contacting the sensor detectors is changing, so there will be some relation between the variance present in the sensor responses and solution components. Changing the solution composition at the detector is most conveniently accomplished by chromatography or possibly by flow injection analysis. Extraction of the identities of the underlying factors responsible for generation of the data is accomplished by principal components analysis [33,34]. The three-dimensional matrix D (sensor channel-time-intensity) of sensor responses obtained from such an experiment is to be partitioned into its component matrices of unknown spectra, G and unknown elution profiles U,

$$D = UG \tag{12.27}$$

by the calculation of the abstract factor scores and loadings of DD^T, namely

$$D = VA \tag{12.28}$$

where V is the matrix of abstract chromatograms, and A is the abstract matrix of responses at the detector array. The matrix V is transformed into U by rotation—in essence, by searching for linear combinations of the eigenvectors in V that give good estimates of the chromatogram. The rotation matrix is determined by iterative comparison of a target elution profile t with an estimated profile $\hat{t}$, obtained from V by the rotation r, where

$$r = (V^T V)^{-1} V^T t = V^T t \qquad (12.29)$$

so that $\hat{t} = Vr$. Once C_u is known, the response matrix S_u is calculated by multiple linear regression.

$$S_u = (C_u{}^T C_u)^{-1} C_u{}^T D \qquad (12.30)$$

The sequence of steps outlined above accomplishes the qualitative analysis of the data matrix D. Quantitation requires calibration data, and either regression of the model component on the data set (target rotation) or systematic removal of variance in the data attributed to that component, as weighted by its concentration (rank annihilation). As with PCR and PLS, factor analysis methods can be used to remove the linear effects of species in a stepwise fashion, once species' identities are known, and model data are available from a calibration step. Like methods based more directly on regression, however, multicollinearities in the data may make it difficult to separate component effects from each other. And, like other factor modeling methods, enough data must be provided to characterize all components, but data must be carefully selected because of the large computational burden imposed by the matrix diagonalization and iterative testing steps. Finally, the inability to obtain quantitative information without the assumptions required for the full-model or the partial-model regression methods discussed above greatly limits this method.

III. ESTIMATION OF SENSOR PRECISION

Two factors which strongly affect the quality of estimates obtained by the chemometric software from the multichannel sensor are calibration quality and sensor drift during and after calibration. Calibration quality determines the adequacy of the model for the data analysis. Drift in a sensor can be an early warning sign of sensor failure. Even when a sensor remains useful, significant drift degrades the quality of the calibration step.

The analysis of calibration model quality is easily accomplished in methods which build a statistical (least-squares) model from calibration data. These methods include PCR, PLS, and ILS. To confirm the quality of the calibration model, cross-validation can be

used. In cross-validation, one part of the data is omitted from the
calibration model, and the remaining part used to predict the
"missing" data. The process is repeated, each time with a different
part of the model omitted. Then, the sum of the squared differences between the predicted and the actual data (the omitted calibration data) is calculated. This quantity, called PRESS because it
represents the PRediction Error Sum of Squares, is a measure of
the predictive qualities of the calibration model. With sufficient calibration data, it is possible to select the calibration model so that
the smallest PRESS is obtained [5,17]. Using the PRESS, it is also
possible to determine the number of eigenvectors to retain in a PCR
or a PLS model, and thus to define the "best" calibration model.
For the more empirical methods, a statistical model is not used directly, and PRESS methods have only recently been demonstrated
[35]. Calibration is usually accomplished by the application of the
multicomponent analysis method to pure components, or to mixtures
with well-defined compositions. By these analyses, the quality of
the relation of response to state is established, although the method
is not nearly as well-based as the PRESS method discussed above.

The estimation of sensor drift is much more difficult than the
estimation of calibration quality. The most straightforward approach
is to measure at appropriate intervals some standard sample, and to
compare results of the analysis over time, forming a kind of time
series. These may be analyzed mathematically using standard methods. Systematic components of that time series represent drift in
the sensor response. Of course, this analysis is usually done after
the fact, and correction of the response for sensor drift cannot
easily be made in real time. An alternative approach is to model
the sensor drift as a random walk process and include it in the calibration model [36—41]. Then, the drift can be estimated along
with the rest of the information contained in the system state, and
the effects of the drift can be removed; one way to ensure the accurate removal of drift is to recalibrate the sensor array when the
precision of the estimated concentrations decreases below some preset limit. In this way, the sensor is calibrated only when necessary, and the drift model is continually readjusted to best reflect
the actual drift properties of the sensor. This approach has been
used with Kalman filter estimation on single-channel detection of
chloride ion by flow injection analysis [38—40]. The abrupt
changes in process properties that may occur between batches in
batch-type sampling can also be considered a form of drift, which
can be modeled and corrected like the more conventional forms
above; the between batch variance component can be significantly
reduced using this approach [36,37].

IV. ENSURING OBSERVABILITY IN BIOPROCESSES

In the traditional least-squares method (CLS) and the newer inverse methods (ILS, PCR, and PLS) time-independent behavior in the estimated quantities is presupposed. The presence of sufficient collinearity in the model may make for enough linear dependence to preclude a solution based on multiple linear regression from the model(s) and the data provided. Linearly dependent systems may still be solved for concentrations if the various component concentration profiles differ over time, and a time-dependent model is created. To solve such time-dependent problems, the model must be defined so that the n concentration values (or other components of the state) can be observed from the measured, multicomponent data, which is true when the matrix formed from the time-dependent model (Eqs. (12.19) and (12.20) $[H^T | F^T H^T | \cdots | (F^T)^{n-1} H^T]$ is nonsingular. These time-dependent problems are most directly solved with Kalman filtering methods; many examples exist in the engineering literature, although few have appeared to date in chemistry [24].

Two factors affecting the observability are the optimality of sensor channels chosen for the analysis, and the optimality of the calibration model. Recursive optimization has been used to generate optimal calibration models from generalized standard additions [42] and to find the optimal wavelengths for a spectroscopic analysis of a multicomponent sample [43]. The recursive optimization methods save much time in finding conditions where data multicollinearity is least (in some cases the number of trials is reduced by 10 [42]). These methods are easily automated, and they may point the way to self-calibrating sensors with a little additional research.

Checking, or even optimizing, system observability does not in itself ensure that the results obtained from the multicomponent analysis are ones useful to the control algorithm, however. Many properties can be related to the sensor response, and there is no requirement in any of the methods discussed above that concentrations must be used [44]. For example, PLS methods have been used to predict flavor of peas from near-infrared spectral measurements, and have related mid-infrared band intensities to bulk properties of borosilicate glasses [19]. Kalman filtering methods have been used to associate enzyme activities with ultraviolet spectra of reaction mixtures [26]. Although there is much promise in expanding the scope of such predictions, care must be taken to ensure that a linear (or at least a linearizable) relation exists between the sensor signal and the property desired, since that linear relation is the basis of the multicomponent analysis methods used to date.

Many of the chemometric methods discussed in this chapter have only recently been applied to chemical problems. In some cases, the limitations and advantages of the data analysis are still being explored. Many of these methods would seem to have a place in the simplification of sensor design, however. The predictability of the mathematical separation of mutually interferring responses, the possibility of automating calibration and even optimization of sensor arrays, and the ease with which quality control can be coupled make chemometric methods worthy of considerable study as a part of sensor units.

ACKNOWLEDGMENTS

This work was supported by the U.S. Department of Energy, Office of Basic Energy Sciences, Division of Chemical Sciences, through grant DE-FG02-86ER13542.

REFERENCES

1. Myers, R. H., Classical and Modern Regression with Applications, Duxbury Press, Boston (1986).
2. Haaland, D. M. and Easterling, R. G., Appl. Spectrosc., 34, 539 (1980).
3. Haaland, D. M. and Easterling, R. G., Appl. Spectrosc., 36, 665 (1982).
4. Haaland, D. M., Easterling, R. G., and Vopica, D. A., Appl. Spectrosc., 39, 73 (1985).
5. Beebe, K. R. and Kowalski, B. R., Anal. Chem., 59, 1007A (1987).
6. Otto, M. and Wegscheider, W., Anal. Chem., 57, 63 (1985).
7. Honigs, D. E., Freelin, J. M., Hieftje, G. M., and Hirschfeld, T. B., Appl. Spectrosc., 37, 491 (1983).
8. Alberti, J. C., Phillips, J. A., Fink, D. J., and Wacasz, F. M., Biotech. Bioeng. Symp., 15, 689 (1986).
9. Gendreau, R. M., Chittur, K. K., Dluhy, R. A., and Hutson, T. B., Am. Biotech. Lab., July, 10 (1987).
10. Brown, C. W. and O'Bremski, R. J., Appl. Spectrosc. Rev., 20, 373 (1984).
11. Brown, C. W., Lynch, P. F., O'Bremski, R. J., and Lavery, D. S., Anal. Chem., 54, 1472 (1982).
12. Brown, C. W., O'Bremski, R. J., and Anderson, P., Appl. Spectrosc., 40, 734 (1986).
13. Brown, C. W., Bump, E. A., and O'Bremski, R. J., Appl. Spectrosc., 40, 1023 (1986).

14. Brown, C. W. and Donahue, S. M., "Multiple Ways for Performing Multicomponent Analysis," Abstracts, Digilab Users Group Meeting (1987), pp. 31–37.
15. Haaland, D. M., Spectroscopy, 2(4), 56 (1987).
16. Lorber, A., Wangen, L., and Kowalski, B. R., J. Chemometrics, 1, 19 (1987).
17. Haaland, D. M. and Thomas, E. V., Anal. Chem., 60, 1193 (1988).
18. Haaland, D. M. and Thomas, E. V., Anal. Chem., 60, 1202 (1988).
19. Haaland, D. M., Anal. Chem., 60, 1208 (1988).
20. Martens, H., Karstang, T., and Naes, T., J. Chemometrics, 1, 201 (1987).
21. Cahn, F. and Compton, S., "FTIR Quantitative Analyses by Principal Component Regression and Partial Least Squares," Abstracts, Digilab Users Group Meeting (1987), p. 24.
22. UNSCRAMBLER, Norsk Institute for Naeringsmiddelforskning, Norway.
23. ARTHUR, Infometrix, Seattle, WA.
24. Gelb, A. (ed.), Applied Optimal Estimation, MIT Press, Cambridge (1974).
25. Brown, S. D., Anal. Chim. Acta, 181, 1 (1986).
26. Rutan, S. C. and Brown, S. D., Anal. Chim. Acta, 175, 219 (1985).
27. Bierman, G. J. and Thornton, C. L., Automatica, 13, 23 (1977).
28. Rutan, S. C. and Brown, S. D., Anal. Chim. Acta, 160, 99 (1984).
29. Rutan, S. C. and Brown, S. D., Anal. Chim. Acta, 167, 39 (1985).
30. Wilk, H. R. and Brown, S. D., J. Chemometrics, in press (1990).
31. Brown, S. D., unpublished results.
32. Otto, M. and Bandmeier, H., Chemometrics and Intell. Lab. Systems, 1, 71 (1986).
33. Gemperline, P. J., J. Chem. Inf. Comp. Sci., 24, 206 (1984).
34. Vandeginste, B. G. M., Leyten, F., Gerritsen, M., Noor, J. W., Kateman, G., and Frank, J., J. Chemometrics, 1, 57 (1987).
35. Ansley, C. F. and Kohn, R., Biometrika, 73, 467 (1986).
36. Poulisse, H. N. J. and Jansen, R. T. P., Anal. Chim. Acta,
37. Jansen, R. T. P. and Poulisse, H. N. J., Anal. Chim. Acta, 151, 441 (1983).
38. Thijssen, P. C., Wolfrum, S. M., Kateman, G., and Smit, H. C., Anal. Chim. Acta, 156, 87 (1984).
39. Thijssen, P. C., Anal. Chim. Acta, 162, 253 (1984).

40. Thijssen, P. C., Kateman, G., and Smit, H. C., Anal. Chim.
     Acta, 173, 265 (1985).
41. Pijpers, F. W., Anal. Chim. Acta, 190, 79 (1986).
42. Thijssen, P. C., Vogels, L. J. P., Smit, H. C., and Kateman,
     G., Z. Anal. Chem., 320, 531 (1985).
43. Vandeginste, B. G. M., Klaessens, J., and Kateman, G.,
     Anal. Chim. Acta, 150, 71 (1983).
44. Astrom, K. J. and Wittenmark, B., Computer-Controlled Sys-
     tems: Theory and Design, Prentice-Hall, Englewood Cliffs
     (1984).

Index

A

Absorbance, 98,102
Acetaldehyde, 79
Acetyl cholinesterase, 216
Acoustic sensor, 273
Adaptive control, 260,266—268
Adaptive filter, 310
Aeration, 76
Affinity chromatography, 40
Affinity sensor, 272
Agitation, 76
Alcaligenes eutrophus, 82
Alcohol dehydrogenase, 88,116, 216
Alcohol oxidase, 115,183,184, 216
American Society for Testing and Materials, 18
Ammonia, 101,107,165,185
Ammonia sensor, 107,113,116, 180,185,262
Amperometric methods, 174
Ampicillin, 215
Antibody, 117,174
Antigen, 117
Apyrase, 205
ARTHUR, 305
Ascorbate oxidase, 215
Ascorbic acid, 186,215
Aseptic conditions, 17,35
ASTM, 18

Attenuated total reflectance, 298
Aureobasidium pullulans, 87
Automated sampling, 37

B

Bacillus cereus, 233,235,236
Bacillus subtilis, 82, 105
Baker's yeast, 7,8,12,79,83,85—87,105,106,109,251,253,262
Ball valve, 28,36
Beirman-Thorton algorithm, 307
Bellman's dynamic programming, 288
Benzoquinone, 178,182
β-lactams, 216
Biomass, 2,3,8,12,76,81,85,87, 93,105,106,255,273,274
Biosensor
  electrochemical, 173—191
  ethanol, 115,116
  fiber optic, 100,114
  glucose, 114,115,178,181,182, 272,281
  lactate, 116
  microbial, 175,180,182,183,185
  optical, 87
  penicillin, 115
  tissue, 180
Biuret, 223,227,228,236
Bovine serum albumin, 118

Printed in the United States
by Baker & Taylor Publisher Services